Sarah Trimmer

The story of the robins

Designed to teach children the proper treatment of animals

Sarah Trimmer

The story of the robins
Designed to teach children the proper treatment of animals

ISBN/EAN: 9783337228613

Printed in Europe, USA, Canada, Australia, Japan

Cover: Foto ©berggeist007 / pixelio.de

More available books at **www.hansebooks.com**

THE STORY OF THE ROBINS.

DESIGNED TO TEACH CHILDREN

The Proper Treatment of Animals.

BY

MRS. TRIMMER.

With Coloured Illustrations.

LONDON:
FREDERICK WARNE AND CO.,
BEDFORD STREET, STRAND.
NEW YORK: SCRIBNER, WELFORD, AND ARMSTRONG.

PREFACE.

For more than eighty years the "Story of the Robins" has delighted the children of Great Britain. It is now offered to them, for the first time, with illustrations printed in colours, which it is hoped, will add an additional charm to a book so long and deservedly popular with the young.

August, 1873.

CONTENTS.

CHAPTER I.
                                                     PAGE

HARRIET AND FREDERICK FEEDING THE BIRDS . . . 1

CHAPTER II.
MRS. BENSON AND HER CHILDREN AT BREAKFAST.—THE ROBINS VENTURE UPON THE TEA-TABLE. . . . 15

CHAPTER III.
THE NESTLINGS FRIGHTENED BY THE GARDENER . . . 29

CHAPTER IV.
JOE THE GARDENER BRINGS NEWS OF THE BIRDS' NEST TO HARRIET AND FREDERICK 39

CHAPTER V.
HARRIET AND FREDERICK VIEWING THE ROBINS' NEST . . 46

CHAPTER VI.

	PAGE
THE YOUNG VISITORS.—THE CRUEL BOY	54

CHAPTER VII.

THE FIRST FLIGHT OF THE NESTLINGS 68

CHAPTER VIII.

FREDERICK DISCOVERS THE YOUNG ROBINS IN THE CURRANT BUSH 79

CHAPTER IX.

THE VISIT TO MRS. ADDIS'S 94

CHAPTER X.

ADVENTURES OF THE LITTLE ROBINS 110

CHAPTER XI.

THE FEATHERED NEIGHBOURS 124

CHAPTER XII.

THE VISIT TO THE FARM 134

CHAPTER XIII.

THE PIGS AND BEES 143

CHAPTER XIV.

	PAGE
FREDERICK VIEWING THE DUCKS AND GEESE	153

CHAPTER XV.

THE AVIARY	172

CHAPTER XVI.

THE OLD ROBINS TAKE LEAVE OF THEIR YOUNG ONES	189
CONCLUSION	207

THE STORY OF THE ROBINS.

CHAPTER I.

HARRIET AND FREDERICK FEEDING THE BIRDS.

In a hole which time had made in a wall covered with ivy, a pair of redbreasts built their nest. No place could have been better chosen for the purpose; it was sheltered from the rain, screened from the wind, and in an orchard belonging to a gentleman who had strictly charged his domestics not to destroy the labours of those little songsters who chose his ground as an asylum.

In this happy retreat, which no idle schoolboy dared to enter, the hen redbreast laid four eggs, and then took her seat upon them, resolving that nothing should tempt her to leave the nest for any length of time till she had hatched her infant brood. Her tender mate every morning took her place while she picked up a hasty breakfast, and often, before he tasted any food himself, cheered her with a song.

At length the day arrived when the happy mother heard the chirping of her little ones; with unexpressible tenderness she spread her maternal wings to cover them, threw out the egg-shells in which they before lay confined, then pressed them to her bosom, and presented them to her mate, who viewed them with rapture, and seated himself by her side that he might share her pleasure.

"We may promise ourselves much delight in rearing our little family," said he, "but it will give us a great deal of trouble. I would willingly bear the whole myself, but it will be impossible for me, with my utmost labour and industry, to supply all our nestlings with what is sufficient for their daily support; it will therefore be necessary for you to leave the nest sometimes to seek provisions for them." She declared her readiness to do so, and said that there would be no necessity for her to be long absent, as she had discovered a place near the orchard where food was scattered on purpose for such birds as would take the pains of seeking it; and that she had been informed by a chaffinch that there was no kind of danger in picking it up.

"This is a lucky discovery indeed for us," replied her mate; "for this great increase of family renders it prudent to make use of every means for supplying our necessities. I myself must take a larger circuit, for some insects that are proper for the nestlings cannot be found in all places; however, I will bear you company whenever it is in my power."

The little ones now began to be hungry, and opened their gaping mouths for food; on which their kind father instantly flew forth to find it for them, and in turns supplied them all, as well as his beloved mate. This was a hard day's work, and when evening came on he was glad to take repose, and turning his head under his wing, he soon fell asleep; his mate soon followed his example. The four little ones had before fallen into a gentle slumber, and perfect quietness for some hours reigned in the nest.

The next morning they were awakened at the dawn of day by the song of a skylark, which had a nest near the orchard; and as the young redbreasts were impatient for food, their father cheerfully prepared himself to renew his toil, requesting his mate to accompany him to the place she had mentioned. "That I will do," replied she, "but it is too early yet;

I must therefore beg that you will go by yourself and procure a breakfast for us, as I am fearful of leaving the nestlings before the air is warmer, lest they should be chilled." To this he readily consented, and fed all his little darlings; to whom, for the sake of distinction, I shall give the names of Robin, Dicky, Flapsy, and Pecksy. When this kind office was performed he perched on a tree, and while he rested, entertained his family with his melody, till his mate, springing from the nest, called him to attend her; on which he instantly took wing, and followed her to a courtyard belonging to a family mansion.

No sooner had the happy pair appeared before the parlour window, than it was hastily thrown up by Harriet Benson, a little girl about eleven years old, the daughter of the gentleman and lady to whom the house belonged. Harriet with great delight called her brother to see two robin redbreasts; and she was soon joined by Frederick, a fine chubby rosy-cheeked boy, about six years of age, who, as soon as he had taken a peep at the feathered strangers, ran to his mamma, and entreated her to give him something to feed them with. "I must have a great piece of bread this morning," said he, "for there are all the sparrows and chaffinches that come every day, and two robin redbreasts besides." "Here is a piece for you, Frederick," replied Mrs. Benson, cutting a loaf that was on the table; "but if your daily pensioners continue to increase as they have done lately, we must provide some other food for them, as it is not right to cut pieces from a loaf on purpose for birds, because there are many children who want bread, to whom we should give the preference. Would you deprive a poor little hungry boy of his breakfast to give it to birds?" "No," said Frederick, "I would sooner give my own breakfast to a poor boy than he should go without; but where shall I get food enough for my birds? I will beg the cook to save the crumbs in the bread-pan, and desire John to preserve all he makes when he cuts the loaf for dinner, and those which are scattered on the tablecloth." "A very good scheme," said Mrs. Benson, "and I make no doubt it will answer your purpose, if you can prevail on the servants to indulge you. I cannot bear to see the least fragment of food wasted which may contribute to the support of life in any creature."

Harriet, being quite impatient to exercise her benevolence, requested her brother to remember that the poor birds, for whom he had been a successful solicitor, would soon fly away if he did not make haste to feed them; on which he ran to the window with his treasure in his hand.

When Harriet first appeared, the winged suppliants approached with eager expectation of the daily handful which their kind benefactress made it a custom to distribute, and were surprised at the delay of her charity. They hopped around the window—they chirped—they twittered, and employed all their little arts to gain attention; and were on the point of departing, when Frederick, breaking a bit from the piece he held in his hand, attempted to scatter it among them, calling out at the same time, "Dicky! Dicky!" On hearing the well-known sound, the little flock immediately drew near. Frederick begged that his sister would let him feed all the birds himself; but finding that he could not fling the crumbs far enough for the redbreasts, who, being strangers, kept at a distance, he resigned the task, and Harriet, with dexterous hand, threw some of them to the very spot where the affectionate pair stood waiting for her notice, who with grateful hearts picked up the portion assigned them; and in the meanwhile the other birds, being satisfied, flew away, and they were left alone. Frederick exclaimed with rapture that the two robin redbreasts were feeding; and Harriet meditated a design of taming them by kindness. "Be sure, my dear brother," said she, "not to forget to ask the cook and John for the crumbs, and do not let the least morsel of anything you have to eat fall to the ground. I will be careful in respect of mine, and we will collect all that papa and mamma crumble; and if we cannot by these means get enough, I will spend some of my money in grain for them." "Oh," said Frederick, " I would give all the money I have in the world to buy food for my dear dear birds." "Hold, my love," said Mrs. Benson "though I commend your humanity, I must remind you again that there are poor people as well as poor birds." "Well, mamma," replied Frederick, "I will only buy a little grain, then." As he spoke these last words, the redbreasts having finished their meal, the mother bird expressed her impatience to return to the nest; and having obtained her mate's consent, she repaired with all possible speed to her humble habitation, whilst he tuned his melodious pipe, and delighted their young benefactors with his music; he then spread his wings, and took his flight to an adjoining garden, where he had a great chance of finding worms for his family.

Frederick expressed great concern that the robins were gone; but was comforted by his sister, who reminded him that in all probability his new favourites, having met with so kind a reception, would return on the morrow. Mrs. Benson then bid them shut the window; and taking Frederick in her lap, and desiring Harriet to sit down by her, thus addressed them:—

"I am delighted, my dear children, with your humane behaviour towards animals, and wish by all means to encourage it; but let me recommend to you not to suffer your tender feelings towards animals to gain upon you to such a degree as to make you unhappy or forgetful of those who have a higher claim to your attention—I mean poor people; always keep in mind the distresses which they endure, and on no account waste any kind of food, nor give to inferior creatures what is designed for mankind."

Harriet promised to follow her mamma's instructions; but Frederick's attention was entirely engaged by watching a butterfly, which had just left the chrysalis, and was fluttering in the window, longing to try its wings in the air and sunshine; this Frederick was very desirous to catch, but his mamma would not permit him to attempt it, because, she told him, he could not well lay hold of its wings without doing it an injury, and it would be much happier at liberty. "Should you like, Frederick," said she, " when you are going out to play, to have anybody lay hold of you violently, scratch you all over, then offer you something to eat which is very disagreeable, and perhaps poisonous, and shut you up in a little dark room? And yet this is the fate to which many a harmless insect is condemned by thoughtless children." As soon as Frederick understood that he could not catch the butterfly without hurting it, he gave up the point, and assured his mamma he did not want to keep it, but only to carry it out of doors. "Well," replied she, "that end may be answered by opening the window;" which, at her desire, was done by Harriet: the happy insect was glad to fly away, and Frederick had soon the pleasure of seeing it upon a rose.

Breakfast being ended, Mrs. Benson reminded the children that it was almost time for their lessons to begin; but desired their maid to take them into the garden before they applied to business. During his walk, Frederick amused himself with watching the butterfly as it flew from flower to flower, which gave him more pleasure than he could possibly have received from catching and confining the little tender creature.

Let us now see what became of our redbreasts after they left their young benefactors.

The hen bird, as I informed you, repaired immediately to the nest; her heart fluttered with apprehension as she entered it, and she eagerly called out, "Are you all safe, my little dears?" "All safe, my good mother," replied Pecksy, "but a little hungry, and very cold." "Well," said she, "your last complaint I can soon remove; but in respect to satisfying your hunger, that must be your father's task; however, he will soon be here, I make no doubt." Then spreading her wings over them all, she soon gave warmth to them, and they were again comfortable.

In a very short time her mate returned; for he only stayed at Mr. Benson's to finish his song and sip some clear water, which his new friends always kept where they fed the birds. He brought in his mouth a worm, which was given to Robin; and was going to fetch one for Dicky, but his mate said, "My young ones are now hatched, and you can keep them warm as well as myself; take my place, therefore, and the next excursion shall be mine." "I consent," answered he, "because I think a little flying now and then will do you good; but, to save you trouble, I can direct you to a spot where you may be certain of finding worms for this morning's supply." He then described the place; and on her quitting the nest he entered it, and gathered his young ones under his wings. "Come, my dears," said he, "let us see what kind of a nurse I can make; but an awkward one, I fear; even every mother bird is not a good nurse, but you are very fortunate in yours, for she is a most tender one, and I hope you will be dutiful for her kindness." They all promised him they would. "Well, then," said he, "I will sing you a song." He did so, and it was a very merry one, and delighted the nestlings extremely; so that, though they were not quite comfortable under his wings, they did not regard it, nor think the time of their mother's absence long. She had not succeeded in the place she first went to, as a boy was picking up worms to angle with, of whom she was afraid, and therefore flew further; but as soon as she had obtained what she went for, she returned with all possible speed; and though she had repeated invitations from several gay birds which she met to join their sportive parties, she kept a steady course, preferring the pleasure of feeding little Dicky to all the diversions of the fields and groves. As soon as the hen bird came near the nest her mate started up to make room for her, and take his turn of providing for his family. "Once more adieu!" said he, and was out of sight in an instant.

"My dear nestlings," said the mother, "how do you do?" "Very well, thank you," replied all at once; "and we have been exceedingly merry," said Robin, "for my father has sung us a sweet song." "I think," said Dicky, "I should like to learn it." "Well," replied the mother, "he will teach it you, I dare say; here he comes, ask him." "I am ashamed," said Dicky. "Then you are a silly bird. Never be ashamed but when you commit a fault; asking your father to teach you to sing is not one; and good parents delight to teach their young ones everything that is proper and useful. Whatever so good a father sets you an example of you may safely desire to imitate." Then addressing herself to her mate, who for an instant stopped at the entrance of the nest, that he might not interrupt her instructions, "Am I not right," said she, "in what I have just told them?" "Perfectly so," replied he; "I shall have pleasure in teaching them all that is in my power; but we must talk of that another time. Who is to feed poor Pecksy?" "Oh, I, I!" answered the mother, and was gone in an instant.

"And so you want to learn to sing, Dicky?" said the father: "well, then, pray listen very attentively; you may learn the notes, though you will not be able to sing till your voice is stronger."

Robin now remarked that the song was very pretty indeed, and expressed his desire to learn it also. "By all means," said his father; "I shall sing it very often, so you may learn it if you please." "For my part," said Flapsy, "I do not think I could have patience to learn it, it will take so much time." "Nothing, my dear Flapsy," answered the father, "can be acquired without patience, and I am sorry to find yours begin to fail you already; but I hope, if you have no taste for music, that you will give the greater application to things that may be of more importance to you." "Well," said Pecksy, "I would apply to music with all my heart, but I do not believe it possible for me to learn it." "Perhaps not," replied her father, "but I do not doubt you will apply to whatever your mother requires of you; and she is an excellent judge both of your talents and of what is suitable to your station in life. She is no songstress herself, and yet she is very clever, I assure you: here she comes." Then rising to make room for her, "Take your seat, my love," said he, " and I will perch upon the ivy." The hen again covered her brood, whilst her mate amused her with his singing and conversation till the evening, excepting that each parent bird flew out in turn to get food for their young ones.

In this manner several days passed with little variation; the nestlings were very thriving, and daily gained strength and knowledge, through the care of their indulgent parents, who every day visited their friends, the little Bensons. Frederick had been successful with the cook and footman, from whom he obtained enough for his dear birds, as he called them, without robbing the poor; and he was still able to produce a penny whenever his papa or mamma pointed out to him a proper object of charity.

CHAPTER II.

MRS. BENSON AND HER CHILDREN AT BREAKFAST.—THE ROBINS VENTURE UPON THE TEA-TABLE.

It happened one day that both the redbreasts, who always went together to Mrs. Benson's (because if one had waited for the other's return, it would have missed the chance of being fed),—it happened, I say, that they were both absent longer than usual, for their little benefactors, having been fatigued with a very long walk the evening before, lay late in bed that morning; but as soon as Frederick was dressed, his sister, who was waiting for him, took him by the hand and led him down-stairs, where he hastily asked the cook for the collection of crumbs. As soon as he entered the breakfast-parlour, he ran eagerly to the window, and attempted to fling it up. "What is the cause of this mighty bustle?" said his mamma;

'do you not perceive that I am in the room, Frederick?" "Oh, my birds! my birds!" cried he. "I understand," rejoined Mrs. Benson, "that you have neglected to feed your little pensioners; how came this about, Harriet?" "We were so tired last night," answered Harriet, "that we overslept ourselves, mamma." "This excuse may satisfy you and your brother," answered the lady, "but I fear your birds would bring heavy complaints against you, were they able to talk. But make haste to feed them now; and for the future, whenever you give any living creature cause to depend on you for sustenance, be careful on no account to disappoint it; and if you are prevented from feeding it yourself, employ another person to do it for you."

"It is customary," continued Mrs. Benson, "for little boys and girls to pay their respects to their papas and mammas every morning, as soon as they see them. This, Frederick, you ought to have done to me on entering the parlour, instead of running across it, crying out, 'My birds! my birds!' it would have taken you but very little time to have done so. However, I will excuse your neglect now, my dear, as you did not intend to offend me; but remember, that you depend as much on your papa and me for eveything you want as these little birds do on you;

nay, more so, for they could find food in other places but children can do nothing towards their own support; they should therefore be dutiful and respectful to those whose tenderness and care they constantly experience."

Harriet promised her mamma that she would on all occasions endeavour to behave as she wished her to do; but I am sorry to say Frederick was more intent on opening the window than imbibing the good instructions that were given him. This he could not do; therefore Harriet, with her mamma's permission, went to his assistance, and the store of provisions was dispensed. As many of the birds had nests, they ate their meal with all possible expedition. Among this number were the robins, who despatched the business as soon as they could, for the hen was anxious to return to her little ones, and the cock to procure them a breakfast; and having given his young friends a song before they left their bedchambers, he did not think it necessary to stay to sing any more; they therefore departed.

When the mother bird arrived at the ivy-wall, she stopped at the entrance of the nest with a palpitating heart; but seeing her brood all safe and well, she hastened to take them under her wings. As soon as she was seated she observed that they were not so cheerful as usual. "What is the matter?" said she; "how have you agreed during my absence?" To these questions all were unwilling to reply; for the truth was that they had been quarrelling almost the whole time. "What! all silent?" said she. "I fear you have not obeyed my commands, but have been contending. I desire you will tell me the truth." Robin, knowing that he was the greatest offender, began to justify himself before the others could have time to accuse him.

"I am sure, mother," said he, "I only gave Dicky a little peck because he crowded me so; and all the others joined with him, and fell upon me at once."

"Since you have begun, Robin," answered Dicky, "I must speak, for you gave me a very hard peck indeed and I was afraid you had put out my eye. I am sure I made all the room I could for you; but you said you ought to have half the nest, and to be master when your father and mother were out, because you are the eldest."

"I do not love to tell tales," said Flapsy, "but what Dicky says is very true, Robin; and you plucked two or three little feathers out of me, only because I begged you not to use us ill."

"And you set your foot very hard upon me," cried Pecksy, "for telling you that you had forgotten your dear mother's command.

"This is a sad story indeed," said the mother. "I am very sorry to find, Robin, that you already display such a turbulent disposition. If you go on in this manner we shall have no peace in the nest, nor can I leave it with any degree of satisfaction. As for your being the eldest, though it makes me show you a preference on all proper occasions, it does not give you a privilege to domineer over your brother and sisters. You are all equally the objects of our tender care, which we shall exercise impartially among you, provided you do not forfeit it by bad behaviour. To show you that you are not master of the nest, I desire you to get from under my wing, and sit on the outside, while I cherish those who are dutiful and good." Robin, greatly mortified, retired from his mother; on which Dicky, with the utmost kindness, began to intercede for him. "Pardon Robin, my dear mother, I entreat you," said he; "I heartily forgive his treatment of me, and would not have complained to you, had it not been necessary for my own justification.

"You are a good bird, Dicky," said his mother, "but such an offence as this must be repented of before it is pardoned." At this instant her mate returned with a fine worm, and looked as usual for Robin, who lay sulking by himself. "Give it," said the mother, "to Dicky; Robin must be served last this morning; nay, I do not know whether I shall permit him to have any food all day." Dicky was very unwilling to mortify his brother; but on his mother's commanding him not to detain his father, he opened his mouth and swallowed the delicious mouthful. "What can be the matter?" said the good father, when he had emptied his mouth; "surely none of the little ones have been naughty? But I cannot stop to inquire at present, for I left another fine worm, which may be gone if I do not make haste back."

As soon as he departed, Dicky renewed his entreaties that Robin might be forgiven; but as he sat swelling with anger and disdain, because he fancied that the eldest should not be shoved to the outside of his mother's wing while the others were fed, she would not hear a word in his behalf. The father soon came and fed Flapsy, and then, thinking it best for his mate to continue her admonitions, he flew off again. During her father's absence, Pecksy, whose little heart was full of affectionate concern for the punishment of her brother, thus attempted to comfort him:

"Dear Robin, do not grieve; I will give you my breakfast, if my mother will let me." "Oh," said Robin, "I do not want any breakfast; if I may not be served first, I will have none." "Shall I ask my mother to forgive you?" said Pecksy. "I do not want any of your intercessions," replied he; "if you had not been a parcel of ill-natured things, I should not have been pushed about as I am."

"Come back, Pecksy," said the mother, who overheard them; "I will not have you converse with so naughty a bird. I forbid every one of you even to go near him." The father then arrived, and Pecksy was fed. "You may rest yourself, my dear," said the mother; "your morning's task is ended." "Why, what has Robin done?" asked he. "What I am sorry to relate," she replied,—"quarrelled with his brother and sisters!" "You surprise me; I could not have suspected he would have been either so foolish or so unkind." "Oh, this is not all," said the mother, "for he presumes on being the eldest, and claims half the nest to himself when we are absent, and now is sullen because he is disgraced, and is not fed first as usual." "If this be the case," replied the father, "leave me to settle this business, my dear, and pray go into the air a little, for you seem to be sadly vexed." "I am disturbed," said she, "I confess; for, after all my care and kindness, I did not expect such a sad return as this. I am sorry to expose this perverse bird even to you, but he will not be corrected by me. I will do as you desire, and go into the open air a little." So saying, she repaired to a neighbouring tree, where she anxiously awaited the result of her mate's admonition.

As soon as the mother departed, the father thus addressed the delinquent :—" And so, Robin, you want to be master of the nest? A pretty master you would make, indeed, who do not know even how to govern your own temper! I will not stand to talk much to you now, but depend upon it, I will not suffer you to use any of the family ill, particularly your good mother; and if you persist in obstinacy, I will certainly turn you out of the nest before you can fly." These threatenings intimidated Robin, and he also began to be very hungry as well as cold; he therefore promised to behave better for the future, and his brother and sisters pleaded earnestly that he might be forgiven and restored to his usual place.

" I can say nothing in respect to the last particular," replied the father; "that depends upon his mother; but as it is his first offence, and he seems to be very sorry, I will myself pardon it, and intercede for him with his mother." On this he left the nest to seek for her. "Return, my dear," said he, "to your beloved family; Robin seems sensible of his offence, and longs to ask your forgiveness." Pleased at this intelligence, the mother raised her drooping head, and closed her wings, which hung mournfully by her sides, expressive of the dejection of her spirits. "I fly to give it him," said she, and hastened into the nest. In the meanwhile Robin wished for, yet dreaded, her return.

As soon as he saw her he lifted up a supplicating eye, and in a weak tone (for hunger and sorrow had made him faint) he cried, "Forgive me, dear mother; I will not again offend you." "I accept your submission, Robin," said she, "and will once more receive you to my wing; but indeed your behaviour has made me very unhappy." She then made room for him, he nestled closely to her side, and soon found the benefit of her fostering heat; but he was still hungry, yet he had not confidence to ask his father to fetch him any food; but this kind parent, seeing that his mother had received him into favour, flew with all speed to an adjacent field, where he soon met with a worm, which with tender love he presented to Robin, who swallowed it with gratitude. Thus was peace restored to the nest, and the happy mother once more rejoiced that harmony reigned in the family.

A few days after, a fresh disturbance took place. All the little redbreasts, excepting Pecksy, in turn committed some fault or other, for which they were occasionally punished; but she was of so amiable a disposition, that it was her constant study to act with propriety, and avoid giving offence; on which account she was justly caressed by her parents with distinguished kindness. This excited the envy of the others, and they joined together to treat her ill, giving her the title of the Favourite; saying that they made no doubt that their father and mother would reserve the nicest morsels for their darling.

Poor Pecksy bore all their reproaches with patience, hoping that she should in time regain their good opinion by her gentleness and affection. But it happened one day that, in the midst of their tauntings, their mother unexpectedly returned, who, hearing an uncommon noise among her young ones, stopped on the ivy to learn the cause, and as soon as she discovered it, she made her appearance at the entrance of the nest, with a countenance that showed she knew what was going on.

"Are these the sentiments," said she, "that subsist in a family which ought to be bound together by love and kindness? Which of you has cause to reproach either your father or me with partiality? Do we not with the exactest equality distribute the fruits of our labours among you? And in what respect has poor Pecksy the preference, but in that praise which is justly her due, and which you do not strive to deserve? Has she ever yet uttered a complaint against you? though, from the dejection of her countenance, which she in vain attempted to conceal, it is evident that she has suffered your reproaches for some days past. I positively command you to treat her otherwise, for it is a mother's duty to succour a persecuted nestling; and I will certainly admit her next my heart, and banish you all from that place you have hitherto possessed in it, if you suffer envy and jealousy to occupy your bosoms, instead of that tender love which she, as the kindest of sisters, has a right to expect from you."

Robin, Dicky, and Flapsy were quite confounded by their mother's reproof; and Pecksy, sorry that they had incurred the displeasure of so tender a parent, kindly endeavoured to soften her anger. "That I have been vexed, my dear mother," said she, " is true, but not so much as you suppose; and I am ready to believe that my dear brothers and sister were not in earnest in the severe things they said of me—perhaps they only meant to try my affection. I now entreat them to believe that I would willingly resign the greatest pleasure in life, could I by that means increase their happiness; and so far from wishing for the nicest morsel, I would content myself with the humblest fare, rather than any of them should be disappointed."

This tender speech had its desired effect; it recalled those sentiments of love which envy and jealousy had for a time banished; all the nestlings acknowedged their faults, their mother forgave them, and a perfect reconciliation took place, to the great joy of Pecksy, and indeed of all parties.

All the nestlings continued very good for several days, and nothing happened worth relating. The little family were soon covered with feathers, which their mother taught them to dress, telling them that neatness was a very essential thing, both for health, and also to render them agreeable to the eye of the world.

Robin was a very strong, robust bird, not remarkable for his beauty, but there was a great briskness in his manner, which covered many defects, and he was very likely to attract notice. His father judged, from the tone of his chirpings, that he would be a very good songster.

Dicky had remarkably fine plumage; his breast was of a beautiful red, his body and wings of an elegant mottled brown, and his eyes sparkled like diamonds.

Flapsy was also very pretty, but more distinguished for the elegance of her shape than for the variety and lustre of her feathers.

Pecksy had no outward charms to recommend her to notice; but these defects were supplied by the sweetness of her disposition. Her temper was constantly serene, she was ever attentive to the happiness of her parents, and would not have grieved them for the world; and her affection for her brothers and sister was so great, that she constantly preferred their interest to her own, of which we have lately given an instance.

The kind parents attended to them with unremitting affection, and made their daily visit to Frederick and Harriet Benson, who very punctually discharged the benevolent office of feeding them. The robin redbreasts, made familiar by repeated favours, approached nearer and nearer to their little friends by degrees and at length ventured to enter the room and feed upon the breakfast-table. Harriet was delighted at this circumstance, and Frederick was quite transported; he longed to catch the birds, but his mamma told him that would be the very means to drive them away. Harriet entreated him not to frighten them on any account, and he was prevailed upon to forbear, but could not help expressing a wish that he had them in a cage, that he might feed them all day long.

"And do you really think, Frederick," said Mrs. Benson, "that these little delicate creatures are such gluttons as to desire to be fed all day long? Could you tempt them to do it, they would soon die; but they know better, and as soon as their appetites are satisfied, always leave off eating. Many a little boy may learn a lesson from them. Do you not recollect one of your acquaintances, who, if an apple-pie or anything that he calls nice is set before him, will eat till he makes himself sick?" Frederick looked ashamed, being conscious that he was too much inclined to indulge his love of delicacies. "Well," said his mamma, "I see you understand who I mean, Frederick, so we will say no more on that subject; only when you meet with that little gentleman, give my love to him, and tell him I beg he will be as moderate as his redbreasts."

CHAPTER III.

THE NESTLINGS FRIGHTENED BY THE GARDENER.

THE cock bird, having finished his breakfast, flew out at the window, followed by his mate; and as soon as they were out of sight, Mrs. Benson continued her discourse:—"And would you really confine these sweet creatures in a cage, Frederick, merely to have the pleasure of feeding them? Should you like to be always shut up in a little room, and think it sufficient if you were supplied with victuals and drink? Is there no enjoyment in running about, jumping, and going from place to place? Do you not like to keep company with little boys and girls? And is there no pleasure in breathing the fresh air? Though these little animals are inferior to you, there is no doubt but they are capable of enjoyments similar to these; and it must be a dreadful life for a poor bird to be shut up in a cage, where he cannot so much as make use of his wings, where he is separated from his natural companions, and where he cannot possibly receive that refreshment which the air must afford to him when at liberty to fly to such a height. But this is not all; for many a poor bird is caught and taken from its family, after it has been at the trouble of building a nest, has perhaps laid its eggs, or even hatched its young ones, which are by this means exposed to certain destruction. It is likely that these very redbreasts may have young ones, for this is the season of the year for their hatching; and I rather think they have from the circumstance of their always coming together."

"If that be the case," said Harriet, "it would be a pity indeed to confine them. But why, mamma, if it is wrong to catch birds, did you at one time keep canary-birds?"

"The case is very different in respect to canary-birds, my dear," said Mrs. Benson; "by keeping them in a cage I did them a kindness. I considered them as little foreigners who claimed my hospitality. This kind of bird came originally from a warm climate; they are in their nature very susceptible of cold, and would perish in the open air in our winters; neither does the food which they feed on grow plentifully in this country; and as here they are always bred in cages, they do not know how to procure the materials for their nest abroad. And there is another particular which would greatly distress them were they to be turned loose, which is the persecution they would be exposed to from other birds. I remember once to have seen a poor hen canary-bird, which had been turned loose because it could not sing; and surely no creature could be more miserable. It was starving for want of food, famishing with thirst, shivering with cold, and looked terrified to the greatest degree; while a parcel of sparrows and chaffinches pursued it from place to place, twittering and chirping with every mark of insult. I could not help fancying the little creature to be like a foreigner just landed from some distant country, followed by a rude rabble of boys, who were ridiculing him because his dress and language were strange to them."

"And what became of the poor little creature, mamma?" said Harriet. "I was going to tell you, my dear," replied Mrs. Benson; "I ordered the servant to bring me a cage, with seed and water in their usual places; this I caused to be hung on a tree, next to that in which the little sufferer in vain endeavoured to hide herself among the leaves from her cruel pursuers. No sooner did the servant retire than the poor little wretch flew to it. I immediately had the cage brought into the parlour, where I experienced great pleasure in observing what happiness the poor creature enjoyed in her deliverance. I kept her some years; but not choosing to confine her in a little cage, I had a large one bought, and procured a companion for her of her own species. I supplied them with materials for building; and from them proceeded a little colony, which grew so numerous that you know I gave them to Mr. Bruce to put in his aviary, where you have seen them enjoying themselves. So now I hope I have fully accounted for having kept canary-birds in a cage."

"You have indeed, mamma," said Harriet.

"I have also," said Mrs. Benson, "occasionally kept larks. In severe winters vast numbers of them come to this country from a colder climate, and many perish. Quantities of them are killed and sold for the spit; and the birdcatchers usually have a great many to sell, and many an idle boy has some to dispose of. I frequently buy them, as you know, Harriet; but as soon as the fine weather returns, I constantly set them at liberty. But come, my dears, prepare for your morning walk, and afterwards let me see you in my dressing-room."

"I wonder," said Frederick, "whether our redbreasts have got a nest? I will watch to-morrow which way they fly, for I should like to see the little ones."

"And what will you do, should you find them out?" said his mamma; "not take the nest, I hope?"

"Why," replied Frederick, "I should like to bring it home, mamma, and put it in a tree near the house; and then I would scatter crumbs for the old ones to feed them with."

"Your design is a kind one," said Mrs. Benson, "but you would greatly distress your little favourites. Many birds, through fear, forsake their nests when they are removed; therefore I desire you to let them alone if you should chance to find them." Harriet then remarked that she thought it very cruel to take birds' nests. "Ah, my dear," said Mrs. Benson, "those who commit such barbarous actions are quite insensible to the distresses they occasion. It is very true that we ought not to indulge so great a degree of pity and tenderness for animals as for those who are more properly our fellow-creatures—I mean men, women, and children; but as every living creature can feel, we should have a constant regard to those feelings, and strive to give happiness rather than inflict misery. But go, my dear, and take your walk." Mrs. Benson then left them, to attend her usual morning employments; and the children, attended by their maid, passed an agreeable half-hour in the garden.

In the meantime the hen redbreast returned to the nest, while her mate took his flight in search of food for his family. When the mother approached the nest, she was surprised at not hearing as usual the chirping of her young ones; and what was her astonishment at seeing them all crowded together, trembling with apprehension! "What is the matter, my nestlings," said she, "that I find you in this terror?" "Oh, my dear mother," cried Robin, who first ventured to raise up his head, "is it you?" Pecksy then revived, and entreated her mother to come into the nest, which she did without delay; and the little tremblers crept under her wings, endeavouring to conceal themselves in this happy retreat.

"What has terrified you in this manner?" said she. "Oh! I do not know," replied Dicky; "but we have seen such a monster as I never beheld before," "A monster, my dear? pray describe it." "I cannot," said Dicky; "it was too frightful to be described." "Frightful indeed!" cried Robin; "but I had a full view of it, and will give the best description I can. We were all sitting peaceably in the nest, and very happy together; Dicky and I were trying to sing, when suddenly we heard a noise against the wall, and presently a great round red face appeared before the nest, with a pair of enormous staring eyes, a very large beak, and below that a wide mouth with two rows of bones, that looked as if they could grind us all to pieces in an instant. About the top of this round face, and round the sides, hung something black, but not like feathers. When the two staring eyes had looked at us for some time, the whole thing disappeared."

"I cannot at all conceive from your description, Robin, what this thing could be," said the mother; "but perhaps it may come again." "Oh! I hope not!" cried Flapsy; "I shall die with fear if it does." "Why so, my love?" said her mother; "has it done you any harm?" "I cannot say it has," replied Flapsy. "Well, then, you do very wrong, my dear, in giving way to such apprehensions. You must strive to get the better of this fearful disposition. When you go abroad in the world you will see many strange objects, and if you are terrified at every appearance which you cannot account for, you will live a most unhappy life. Endeavour to be good, and then you need not fear anything. But here comes your father; perhaps he will be able to explain the appearance which has so alarmed you to-day."

As soon as the father had given the worm to Robin, he was preparing to depart for another, but, to his surprise, all the rest of the nestlings begged him to stay, declaring they had rather go without their meal, on condition he would but remain at home and take care of them. "Stay at home and take care of you!" said he; "why, is that more necessary now than usual?" The mother then related the strange occurrence which had occasioned this request. "Nonsense!" said he; "a monster! great eyes! large mouth! long beak! I don't understand such stuff. Besides, as it did them no harm, why are they to be in such terror now it is gone?" "Don't be angry, dear father," said Pecksy, "for it was very frightful indeed." "Well," said he, "I will fly all around the orchard, and perhaps I may meet this monster." "Oh, it will eat you up! it will eat you up!" said Flapsy. "Never fear," said he; and away he flew.

The mother then again attempted to calm them, but all in vain; their fears were now redoubled for their father's safety; however, to their great joy, he soon returned. "Well," said he, "I have seen this monster." The little ones then clung to their mother, fearing the dreadful creature was just at hand.

"What, afraid again?" cried he; "a parcel of stout hearts I have in my nest, truly! Why, when you fly about in the world, you will in all probability see hundreds of such monsters, as you call them, unless you choose to confine yourselves to a retired life; nay, even in woods and groves you will be liable to meet some of them, and those of the most mischievous kind." "I begin to comprehend," said the mother, "that these dear nestlings have seen the face of a man." "Even so," replied her mate; "it is a man, no other than our friend the gardener, who has so alarmed them."

"A man!" cried Dicky; "was that frightful thing a man?" "Nothing more, I assure you," answered his father, "and a good man too, I have reason to believe; for he is very careful not to frighten your mother and me when we are picking up worms, and has frequently thrown crumbs to us when he was eating his breakfast." "And does he live in this garden?" said Flapsy. "He works here very often," replied her father, "but is frequently absent." "Oh, then," cried she, "pray take us abroad when he is away, for indeed I cannot bear to see him." "You are a little simpleton," said the father, "and if you do not endeavour to get more resolution, I will leave you in the nest by yourself when I am teaching your brothers and sister to fly and peck; and what will you do then? for you must not expect we shall go from them to bring you food."

Flapsy, fearful that her father would be quite angry, promised to follow his direction in every respect; and the rest, animated by his discourse, began to recover their spirits.

CHAPTER IV.

JOE THE GARDENER BRINGS NEWS OF THE BIRDS' NEST TO HARRIET AND FREDERICK.

WHILST the terrible commotions related in the last chapter passed in the nest, the monster, who was no other than honest Joe the gardener, went to the house and inquired for his young master and mistress, having, as he justly supposed, some very pleasing news to tell them. Both the young gentleman and lady very readily attended, thinking he had got some fruit or flowers for them. "Well, Joe," said Miss Benson, "what have you to say to us? Have you got a peach or a nectarine, or have you brought me a root of sweet-william?"

"No, Miss Harriet," said Joe; "but I have something to tell you that will please you as much as though I had." "What's that? what's that?"

said Frederick. "Why, Master Frederick," said Joe, "a pair of robins have comed mortal often to one place in the orchard lately; so thinks I, these birds have got a nest. So I watches, and watches, and at last I see'd the old hen fly into a hole in the ivy-wall. I had a fancy to set my ladder and look in; but as master ordered me not to frighten the birds, I stayed till the old one flew out again, and then I mounted, and there I see'd the little creatures full fledged; and if you and Miss Harriet may go with me, I will show them to you, for the nest is but a little way from the ground, and you may easily get up the stepladder."

Frederick was in raptures, being confident that these were the identical robins he was so attached to; and, like a little thoughtless boy as he was, he would have gone immediately with the gardener, had not his sister reminded him that it was proper to ask their mamma's leave first; she therefore told Joe she would let him know when she had done so.

When the redbreasts had quieted the fears of their young family, and fed them as usual, they retired to a tree, desiring their little nestlings not to be terrified if the monster should look in upon them again, as it was very probable he would do.

They promised to bear the sight as well as they could.

When the old ones were seated in the tree, "It is time," said the father, "to take our nestlings abroad. You see, my love, how very timorous they are; and if we do not use them a little to the world, they will never be able to shift for themselves." "Very true," replied the mother; "they are now well fledged, and therefore, if you please, we will take them out to-morrow; but prepare them for it." "One of the best preparatives," answered her mate, "will be to leave them by themselves a little; therefore we will now take a flight together, and then go back." The mother complied, but she longed to be with her dear family.

When they stopped a little to rest on a tree, "Last year," said the hen redbreast, "it was my misfortune to be deprived of my nestlings by some cruel boys, before they were quite fledged, and it is that which makes me so timid now, that I do not feel comfortable when I am away from them."

"A calamity of the same kind befell me," replied the father; "I never shall forget it. I had been taking a flight in the woods in order to procure some nice morsels for one of my nestlings; when I returned to the place in which I had imprudently built.

The first circumstance that alarmed me was a part of my nest scattered on the ground just at the entrance of my habitation; I then perceived a large opening in the wall, where before there was only room for myself to pass. I stopped with a beating heart, in hopes of hearing the chirpings of my beloved family, but all was silent. I then resolved to enter: but what was my consternation when I found that the nest which my dear mate and I had with so much labour built, and the dear little ones who were the joy of our lives, were stolen away! nay, I did not know but the tender mother also was taken. I rushed out of the place distracted with apprehensions for the miseries they might endure, and lamenting my weakness, which rendered me incapable of rescuing them. I was ready to tear off my own feathers with vexation; but recollecting that my dear mate might in all probability have escaped, I resolved to go in search of her.

"As I was flying along I saw three boys, whose appearance was far from disagreeable; one of them held in his hand my nest of young ones, which he eyed with cruel delight, while his companions seemed to share his joy. The dear little creatures, insensible of their fate (for they were newly hatched), opened their mouths, expecting to be fed by me or their mother, but all in vain; to have attempted feeding them at this time would have been inevitable destruction to myself; but I resolved to follow the barbarians, that I might at least see to what place my darlings were consigned.

"In a short time the party arrived at a house, and he who before held the nest now committed it to the care of another, but soon returned with a kind of victuals I was totally unacquainted with, and with this my young ones, when they gaped for food, were fed; hunger induced them to swallow it, but soon after, missing the warmth of their mother, they set up a general cry, which pierced my very heart. Immediately after this the nest was carried away, and what became of my nestlings afterwards I never could discover, though I frequently hovered about the fatal spot of their imprisonment with the hope of seeing them."

"Pray," said the hen redbreast, "what became of your mate?" "Why, my dear," said he, "when I found there was no chance of assisting my little ones, I pursued my course, and sought her in every place of our usual resort, but to no purpose; at length I returned to the bush, where I beheld an afflicting sight indeed—my beloved companion lying on the ground, just expiring! I flew to her instantly, and endeavoured to recall her to life. At the sound of my voice she lifted up her languid eyelids, and said, 'Are you then safe, my love? what is become of our little ones?' In hopes of comforting her, I told her they were alive and well; but she replied, 'Your consolations come too late; the blow is struck, I feel my death approaching. The horror which seized me when I missed my nestlings, and supposed myself robbed at once of my mate and infants, was too powerful for my weak frame to sustain. Oh! why will the human race be so wantonly cruel? The agonies of death now came on, and after a few convulsive pangs she breathed her last, and left me an unhappy widower. I passed the remainder of the summer, and a dreary winter that succeeded it, in a very uncomfortable manner, though the natural cheerfulness of my disposition did not leave me long a prey to unavailing sorrow. I resolved the following spring to seek another mate, and had the good fortune to meet with you, whose amiable disposition has renewed my happiness. And now, my dear," said he, "let me ask you what became of your former companion?"

"Why," replied the hen redbreast, "soon after the loss of our nest, as he was endeavouring to discover what was become of it, a cruel hawk caught him up, and devoured him in an instant. I need not say that I felt the bitterest pangs for his loss; it is sufficient to inform you that I led a solitary life till I met with you, whose endearing behaviour has made society again agreeable to me."

CHAPTER V.

HARRIET AND FREDERICK VIEWING THE ROBINS' NEST.

As soon as Mrs. Benson returned to her children, Frederick ran up to her, saying, "Good news! good news, mamma! Joe has found the robins' nest!" "Has he indeed?" said Mrs. Benson. "Yes, mamma," said Harriet, "and if agreeable to you, we shall be glad to go along with Joe to see it." "But how are you to get at it?" said the lady, "for I suppose it is some height from the ground." "Oh I can climb a ladder very well," cried Frederick. "You climb a ladder! You are a clever gentleman at climbing, I know," replied his mamma; "but do you propose to mount too, Harriet? I think this is rather an indelicate scheme for a lady." "Joe tells me that the nest is but a very little way from the ground, mamma," answered Harriet; "but if I find it otherwise, you may depend on my not going up." "On this condition I will permit you to go," said Mrs. Benson; "but pray, Frederick, let me remind you not to frighten your little favourites." "Not for all the world!" said Frederick. So away he skipped, and ran to Joe before his sister. "We may go! we may go, Joe!" cried he. "Stay for me, Joe, I beg," said Harriet, who presently joined him. Frederick's impatience was so great that he could scarcely be restrained from running all the way, but his sister entreated him not to make himself too hot.

At length they arrived at the desired spot; Joe placed the ladder, and his young master, with a little assistance, mounted it very nimbly; but who can describe his raptures when he beheld the nestlings! "Oh the sweet creatures!" cried he, "there are four of them, I declare! I never saw anything so pretty in my life! I wish I might carry you all home!" "That you must not do, Frederick," said his sister; "and I beg you will come away, for you will either terrify the little creatures or alarm the old birds, which perhaps are now waiting somewhere near to feed them." "Well, I will come away directly," said Frederick; "and so good-bye, robins! I hope you will come soon, along with your father and mother, to be fed in the parlour." He then, under the conduct of his friend Joe, descended.

Joe next addressed Miss Harriet: "Now, my young mistress," said he, "will you go up?" As the steps of the ladder were broad, and the nest was not high, Miss Benson ventured to go up, and was equally delighted with her brother, but so fearful of terrifying the little birds and alarming the old ones, that she would only indulge herself with a peep at the nest. Frederick inquired how she liked the young robins. "They are sweet creatures," said she, "and I hope they will soon join our party of birds, for they appear to me ready to fly. But let us return to mamma, for you know we promised her to stay but a little while; besides, we hinder Joe from his work." "Never mind that," said the honest fellow; "master won't be angry, I'm sartain; and if I thought he would, I would work an hour later to fetch up lost time." "Thank you, Joe," replied Harriet, "but I am sure papa would not desire you to do so."

At this instant Frederick perceived the two redbreasts, who were returning from their proposed excursion, and called to his sister to observe them. He was very desirous to watch whether they would go back to their nest, but she would on no account consent to stay, lest her mamma should be displeased, and lest the birds should be frightened; Frederick, therefore, with reluctance followed her, and Joe attended them to the house.

As soon as they were out of sight the hen bird proposed to return to the nest; she had observed the party, and though she did not see them looking into her habitation, she supposed, from their being so near, that they had been taking a view of it, and told her suspicions to her mate. He agreed with her, and said he now expected to hear a fine story from the nestlings. "Let us return, however," said the mother, "for perhaps they have been terrified again." "Well," said he, "I will attend you then: but let me caution you, my dear, not to indulge their fearful disposition, because such indulgence will certainly prove injurious to them." "I will do the best I can," replied she, and then flew to the nest, followed by her mate.

She alighted upon the ivy, and peeping into the nest, inquired how they all did. "Very well, dear mother," said Robin. "What!" cried the father, who now alighted, "all safe? not one eaten up by the monster?" "No, father," replied Dicky, "we are not devoured; and yet, I assure you, the monster we saw before has been here again, and brought two others with him." "Two others! what, like him-

E

self?" said the father: "I thought, Flapsy, you were to die with apprehension if you saw him again?"

"And so I believe I should have done, had not you, my good father, taught me to conquer my fears," replied Flapsy. "When I saw the top of him, my heart began to flutter to such a degree that I was ready to die, and every feather of me shook; but when I found he stayed but a very little while, I recovered, and was in hopes he was quite gone. My brothers and sisters, I believe, felt as I did; but we comforted one another that the danger was over for this day, and all agreed to make ourselves happy, and not fear this monster, since you assured us he was very harmless. However, before we were perfectly come to ourselves we heard very uncommon noises, sometimes a hoarse sound, disagreeable to our ears as the croaking of a raven, and sometimes a shriller noise, quite unlike the note of any bird that we know of; and immediately after something presented itself to our view which bore a little resemblance to the monster, but by no means so large and frightful. Instead of being all over red, it had on each side two spots of a more beautiful hue than Dicky's breast; the rest of it was of a more delicate white, excepting two streaks of a deep red, like the cherry you brought us the other day, and between these two streaks were rows of white bones, but by no means dreadful to behold, like those of the great monster. Its eyes were blue and white; and round this agreeable face was something which I cannot describe, very pretty, and as glossy as the feathers of a goldfinch. There was so cheerful and pleasing a look in this creature altogether, that, notwithstanding I own I was rather afraid, yet I had pleasure in looking at it; but it stayed a very little time, and then disappeared. While we were puzzling ourselves with conjectures concerning it, another creature, larger than it, appeared before us, equally beautiful, and with an aspect so mild and gentle that we were all charmed with it; but, as if fearful of alarming us by its stay, it immediately retired, and we have been longing for you and my mother's return, in hopes you would be able to tell us what it is we have seen."

"I am happy, my dears," said their mother, "to find you more composed than I expected; for as your father and I were flying together, in order to come back to you, we observed the monster and the two pretty creatures Flapsy has described; the former is, as your father before informed you, our friend the gardener, and the others are our young benefactors, by whose bounty we are every day regaled, and who, I will venture to say, will do you no harm. You cannot think how kindly they treat us; and though there are a number of other birds who share their goodness, your father and I are favoured with their particular regard."

"Oh!" said Pecksy, "are these sweet creatures your friends? I long to go abroad that I may see them again." "Well," cried Flapsy, "I perceive that if we judge from appearances we may often be mistaken. Who would have thought that such an ugly monster as that gardener could have had a tender heart?" "Very true," replied the mother; you must make it a rule, Flapsy, to judge of mankind by their actions, and not by their looks. I have known some of them whose appearance was as engaging as that of our young benefactors, who were, notwithstanding, barbarous enough to take eggs out of a nest and spoil them; nay, even to carry away nest and all before the young ones were fledged, without knowing how to feed them, or having any regard to the sorrows of the tender parents."

"Oh, what dangers there are in the world!" cried Pecksy; "I shall be afraid to leave the nest." "Why so, my love?" said the mother; "every bird does not meet with hawks and cruel children. You have already, as you sat on the nest, seen thousands of the feathered race, of one kind or other, making their airy excursions, full of mirth and gaiety. This orchard constantly resounds with the melody of those who chant from their songs of joy; and I believe there are no beings in the world happier than birds, for we are naturally formed for cheerfulness; and I trust that a prudent precaution, and following the rules we shall from our experience be able to give you, will preserve you from the dangers to which the feathered race are exposed."

"Instead of indulging your fears, Pecksy," said the father, "summon up all your courage, for to-morrow you shall, with your brothers and sisters, begin to see the world."

Dicky expressed great delight at this declaration, and Robin boasted that he had not the least remains of fear. Flapsy, though still apprehensive of monsters, yet longed to see the gaieties of life, and Pecksy wished to comply with every desire of her dear parents. The approach of evening now reminded them that it was time to take repose, and turning its head under its wing, each bird soon resigned itself to the gentle powers of sleep.

CHAPTER VI.

THE YOUNG VISITORS.—THE CRUEL BOY.

AFTER Harriet and Frederick had been gratified with the sight of the robins' nest, they were returning to the house, conducted by their friend Joe, when they were met in the garden by their mamma, accompanied by Miss Lucy Jenkins and her brother Edward. The former was a fine girl about ten years old, the latter a robust, rude boy, more than eleven. "We were coming to seek you, my dears," said Mrs. Benson to her children, "for I was fearful that the business you went upon would make you forgetful of your young visitors."

"I cannot answer for Frederick," replied Harriet, "but indeed, mamma, I would not on any account have slighted my friends.—How do you do, my dear Lucy?" said she; "I am happy to see you. Will you go with me into the play-room? I have got some very pretty new books.—Frederick, have you nothing to show Edward?" "Oh yes," said Frederick, "I have got a new ball, a new top, a new organ, and twenty pretty things; but I had rather go back and show him the robins."

"The robins?" said Edward, "what robins?"

"Why, our robins, that have built in the ivy-wall. You never saw anything so pretty in your life as the little ones."

"Oh, I can see birds enough at home," said Edward; "but why did you not take the nest? it would have been nice diversion to you to toss the young birds about. I have had a great many nests this year, and do believe I have a hundred eggs."

"A hundred eggs! and how do you propose to hatch them?" said Harriet, who turned back on hearing him talk in this manner.

"Hatch them, Miss Benson?" said he; "who ever thinks of hatching birds' eggs?"

"Oh, then, you eat them," said Frederick, "or perhaps let your cook make puddings of them?"

"No, indeed," replied Edward; "I blow out the inside, and then run a thread through them, and give them to Lucy to hang up among her curiosities; and very pretty they look, I assure you."

"And so," said Harriet, "you had rather see a string of empty egg-shells than hear a sweet concert of birds singing in the trees? I admire your taste, truly!"

"Why, is there any harm in taking birds' eggs?" said Lucy; "I never before heard that there was."

"My dear mamma," replied Harriet, "has taught me to think there is harm in every action which gives causeless pain to any living creature; and I own I have a very particular affection for birds."

"Well," said Lucy, "I have no notion of such affections, for my part. Sometimes, indeed, I try to rear those which Edward brings home, but they are teasing, troublesome things, and I am not lucky. To tell the truth, I do not concern myself much about them: if they live, they live; and if they die they die. He has brought me three nests this day to plague me; I intended to have fed the birds before I came out, but being in a hurry to come to see you, I quite forgot it. Did you feed them, Edward?"

"Not I," said he, "I thought you would do it 'tis enough for me to find the nests."

"And have you actually left three nests of young birds at home without food?" exclaimed Harriet.

"I did not think of them, but will feed them when I return," said Lucy.

"Oh!" cried Harriet, "I cannot bear the thought of what the poor little creatures must suffer."

"Well," said Edward, "since you feel so much for them, I think, Harriet, you will make the best nurse. What say you, Lucy, will you give the nests to Harriet?"

"With all my heart," replied his sister; "and pray do not plague me with any more of them."

"I do not know that my mamma will let me accept them," said Harriet; "but if she will, I shall be glad to do so."

Frederick inquired what birds they were, and Edward informed him there was a nest of linnets, a nest of sparrows, and another of blackbirds. Frederick was all impatience to see them, and Harriet longed to have the little creatures in her possession, that she might rescue them from their deplorable condition, and lessen the evils of captivity which they now suffered.

Her mamma had left her with her young companions, that they might indulge themselves in innocent amusements without restraint; but the tenderhearted Harriet could not engage in any play till she had made intercession in behalf of the poor birds; she therefore begged Lucy would accompany her to her mamma, in order to ask permission to have the birds' nests. She accordingly went and made her request known to Mrs. Benson, who readily consented; observing that though she had a very great objection to her children having birds' nests, yet she could not deny her daughter on the present occasion. Harriet, from an unwillingness to expose her friend, had said but little on the subject; but Mrs. Benson, having great discernment, concluded that she made the request from a merciful motive; and knowing that Lucy had no kind mamma to give her instruction, she thus addressed her:—

"I perceive, my young friend, that Harriet is apprehensive that the birds will not meet with the same kind treatment from you which she is disposed to give them. I cannot think you have any cruelty in your nature, but perhaps you have accustomed yourself to consider birds only as playthings, without sense or feeling; to me, who am a great admirer of the beautiful little creatures, they appear in a very different light; and I have been an attentive observer of them, I assure you. Though they have not the gift of speech, like us, all kinds of birds have particular notes, which answer in some measure the purpose of words among them, by means of which they can call to their young ones, express their love for them, their fears for their safety, their anger towards those who would hurt them, &c.; from which we may infer that it is cruel to rob birds of their young, deprive them of their liberty, or exclude them from the blessings suited to their natures, for which it is impossible for us to give them an equivalent. Besides, these creatures, insignificant as they appear in your estimation, were made by God as well as you. Have you not read in the New Testament, my dear, that our Saviour said, 'Blessed are the merciful: for they shall obtain mercy'? How then can you expect that God will send His blessing upon you if, instead of endeavouring to imitate Him in being merciful to the utmost of your power, you are wantonly cruel to innocent creatures which He designed for happiness?"

This admonition from Mrs. Benson, which Lucy did not expect, made her look very serious, and brought tears into her eyes; on which the good lady took her by the hand, and kindly said, "I wish not to distress you, my dear, but merely to awaken the natural sentiments of your heart: reflect at your leisure on what I have taken the liberty of saying to you, and I am sure you will think me your friend. I knew your dear mamma, and can assure you she was remarkable for the tenderness of her disposition. But let me not detain you from your amusements;

go to your own apartment, Harriet, and use your best endeavours to make your visitors happy. You cannot this evening fetch the birds, because when Lucy goes it will be too late for you to take so long a walk, as you must come back afterwards; and I make no doubt but that, to oblige you, she will feed them to-night."

Harriet and Lucy returned, and found Frederick diverting himself with the hand-organ, which had lately been presented to him by his godpapa; but Edward had laid hold of Harriet's dog, and was searching his pocket for a piece of string, that he might tie him and the cat together, to see, as he said, how nicely they would fight; and so fully was he bent on this cruel purpose, that it was with difficulty he was prevailed on to relinquish it.

"Dear me!" said he, "if ever I came into such a house in my life! there is no fun here. What would you have said to Harry Pritchard and me the other day when we made the cats fly?"

"Made the cats fly!" said Frederick; "how was that?"

"Why," replied he, "we tied bladders to each side of their necks, and then flung them from the top of the house. There was an end of their purring and mewing for some time, I assure you, for they lay a long while struggling and gasping for breath, and if they had not had nine lives, I think they must have died; but at last up they jumped, and away they ran scampering. Then out came little Jemmy, crying as if he had flown down himself, because we hurt the poor cats. He had a dog running after him, who, I suppose, meant to call us to task with his bow-wow; but we soon stopped his tongue, for we caught the gentleman, and drove him before us into a narrow lane, and then ran hooting after him into the village; a number of boys joined us, and cried out as we did, 'A mad dog! a mad dog!' On this, several people pursued him with cudgels and broomsticks, and at last he was shot by a man, but not killed, so others came and knocked him about the head till he expired."

"For shame, Edward!" said Harriet; "how can you talk in that rhodomontade manner? I cannot believe any boy could bring his heart to such barbarities."

"Barbarities, indeed! why, have we not a right to do as we please to dogs and cats, or do you think they feel as we do? Fiddle-faddle of your nonsense! say I. Come, you must hear the end of my story: when the dog was dead, we carried him home to little Jemmy, who was ready to break his heart for the loss of him; so we did not like to stand hearing his whining, therefore left him and got a cock, whose legs we tied, and flung at him till he died. Then we set two others fighting; and fine sport we had, for one was pecked till his breast was laid open, and the other was blinded, so we left them to make up their quarrel as they could."

"Stop! stop!" exclaimed Harriet, "for pity's sake, stop! I can hear no more of your horrid stories; nor would I commit even one of those barbarities which you boast of for the world! Poor innocent creatures! what had they done to you to deserve such usage?"

"I beg, Edward," said his sister, "that you will find some other way to entertain us, or I shall really tell Mrs. Benson of you."

"What! are you growing tender-hearted all at once?" cried he.

"I will tell you what I think when I go home," replied Lucy.

As for poor Frederick, he could not restrain his tears, and Harriet's flowed also at the bare idea of the sufferings of the poor animals; but Edward was so accustomed to be guilty of those things without reflection, that there was no making any impression of tenderness upon his mind; and he only laughed at their concern, and wanted to tell a long story about an ox that had been driven by a cruel drover till he went mad; but Harriet and his sister stopped their ears.

At last little Frederick went crying to his mamma, and the young ladies retired to another apartment; so Edward amused himself with catching flies in the window, pulling the legs off some, and the wings off others, delighted with their contortions, which were occasioned by the agonies they endured. Mrs. Benson had some visitors, which prevented her talking to this cruel boy as she otherwise would have done on hearing Frederick's account of him; but she determined to tell his papa, which she accordingly did some time after, when he returned home.

Edward was now disturbed from his barbarous sport by being called to tea; and soon after that was over, the servant came to fetch him and his sister. Harriet earnestly entreated her friend Lucy to feed the birds properly till she should be allowed to fetch them; Lucy promised to do so, for she was greatly affected with Mrs. Benson's discourse, and then entreated her brother to take leave, that she might return home. With this he readily complied, as there were no further opportunities for cruelty.

After her little visitors had departed, Harriet went into the drawing-room, and sat herself down, that she might improve her mind by the conversation of the company. Her mamma perceived that she had been in tears, of which Frederick had before explained the cause. "I do not wonder, my love," said she, "that you should have been so affected with the relation of such horrid barbarities as that thoughtless boy has, by degrees, brought himself to practise by way of amusement. However, do not suffer your mind to dwell on them, as the creatures on which he inflicted them are no longer objects of pity. It is wrong to grieve for the death of animals as we do for the loss of our friends, because they certainly are not of so much consequence to our happiness, and we are taught to think their sufferings end with their lives, as they are not accountable beings; and therefore the killing them, even in the most barbarous manner, is not like murdering a human creature, who is perhaps unprepared to give an account of himself at the tribunal of heaven."

"I have been," said a lady who was present, "for a long time accustomed to consider animals as mere machines, actuated by the unerring hand of Providence to do those things which are necessary for the preservation of themselves and their offspring; but the sight of the Learned Pig, which has lately been shown in London, has deranged these ideas, and I know not what to think."

This led to a conversation on the instinct of animals, which young readers would not understand; it would therefore be useless to insert it.

As soon as the company was gone, "Pray, mamma," said Harriet, "what did the Learned Pig do? I had a great mind to ask Mrs. Franks, who said she saw it; but I was fearful she would think me impertinent."

"I commend your modesty, my dear," replied Mrs. Benson, "but would not have it lead you into such a degree of restraint as to prevent you satisfying that laudable curiosity, without which young persons must remain ignorant of many things very proper for them to be acquainted with. Mrs. Franks would, I am sure, have been far from thinking you impertinent. Those inquiries only are thought troublesome by which children interrupt conversation, and endeavour to attract attention to their own insignificant prattle; but all people of good sense and good nature delight in giving them useful information.

"In respect to the Learned Pig I have heard things which are quite astonishing in a species of animals generally regarded as very stupid. The creature was shown for a sight in a room provided for the purpose, where a number of people assembled to view his performances. Two alphabets of large letters on card-paper were placed on the floor; one of the company was then desired to propose a word which he wished the pig to spell; this the keeper repeated to the pig, which picked out every letter successively with his snout, and collected them together till the word was complete. He was then desired to tell the hour of the day, and one of the company held a watch to him; this he seemed to examine very attentively with his cunning little eye, and having done so, he picked out figures for the hour and minute of the day. He exhibited a number of other tricks of the same nature, to the great diversion of the spectators.

"For my own part, though I was in London at the time he was shown, and heard continually of this wonderful pig from persons of my acquaintance, I never went to see him; for I am fully persuaded that great cruelty must have been used in teaching him things so foreign to his nature, and therefore would not give encouragement to such a scheme."

"And do you think, mamma," said Harriet, "that the pig knew the letters, and could spell words?"

"I think it possible, my dear, that the pig might be taught to know the letters at sight one from the other, and that his keeper had some private sign, by which he directed him to each that was wanted; but that he had an idea of spelling I can never believe, nor are animals capable of attaining human sciences, because for these human faculties are requisite; and no art of man can change the nature of anything, though he may be able to improve that nature to a certain degree, or at least to call forth to view powers which would otherwise be hidden from us. As far as this can be done consistently with our higher obligations, it may be an agreeable amusement, but will never answer any important purpose to mankind; and I would advise you, Harriet, never to give countenance to those people who show what they call learned animals, as you may assure yourself they practise great barbarities upon them, of which starving them almost to death is most likely among the number; and you may, with the money such a sight would cost you, procure for yourself a rational amusement, or even relieve some wretched creature from extreme distress. But, my dear, it is now time for you to retire to rest; I will therefore bid you good-night."

CHAPTER VII.

THE FIRST FLIGHT OF THE NESTLINGS.

EARLY in the morning the hen redbreast awakened her young brood. "Come, my little ones," said she, "shake off your drowsiness; remember, this is the day fixed for your entrance into the world. I desire that each of you will dress your feathers before you go out, for a slovenly bird is my aversion, and neatness is a great advantage to the appearance of every one."

The father bird was upon the wing betimes, that he might give each of his young ones a breakfast before they attempted to leave the nest. When he had fed them he desired his mate to accompany him as usual to Mrs. Benson's, where he found the parlour window open, and his young friends sitting with their mamma. Crumbs had been, according to custom, strewed before the window, which the other birds had nearly devoured; but the redbreasts took their usual post on the tea-table, and the father bird sang his morning lay; after which they returned with all possible speed to the nest, for, having so important an affair to manage, they could not be long absent. Neither could their young benefactors pay so much attention to them as usual, for they were impatient to fetch the birds from Miss Jenkins's; therefore, as soon as breakfast was ended, they set out upon their expedition. Harriet carried a basket large enough to hold two nests, and Frederick a smaller one for the other: thus equipped, with a servant attending them, they set off.

Mr. Jenkins's house was about a mile from Mr. Benson's; it was delightfully situated; there was a beautiful lawn and canal before it, and a charming garden behind; on one side were corn-fields, and on the other a wood. In such a retreat as this it was natural to expect to find a great many birds; but to Harriet's surprise, they saw only a few straggling ones here and there, which flew away the moment she and her brother appeared. On this Harriet observed to Frederick that she supposed Edward Jenkins's practice of taking birds' nests had made them so shy. She said a great deal to him about the cruelties which that naughty boy had boasted of the evening before, which Frederick promised to remember.

As soon as they arrived at the house, Lucy ran out to receive them, but her brother had gone to school.

"We are come, my dear Lucy," said Harriet, 'to fetch the birds you promised us."

"Oh, I know not what to say to you, my dear," said Lucy. "I have very bad news to tell you, and I fear you will blame me exceedingly, though not more than I blame myself. I heartily wish I had returned home immediately after the kind lecture your mamma favoured me with yesterday, which showed me the cruelty of my behaviour, though I was then ashamed to own it. I walked as fast as I could all the way from your house, and determined to give each of the little creatures a good supper, for which purpose I had an egg boiled and nicely chopped; I mixed up some bread and water very smooth, and put a little seed with the chopped egg amongst it, and then carried it to the room where I left the nests. But what was my concern when I found that my care was too late for the greatest part of them! Every sparrow lay dead; they seemed to have killed each other. In the nest of linnets, which were very young, I found one dead, two just expiring, and the other almost exhausted, but still able to swallow; to him, therefore, I immediately gave some of the food I had prepared, which greatly revived him; and as I thought he would suffer with cold in the nest by himself, I covered him over with wool, and had this morning the pleasure of finding him quite recovered."

"What, all the sparrows and three linnets dead!" said Frederick, whose little eyes swam with tears at the melancholy tale; "and pray, Miss Jenkins, have you starved all the blackbirds too?"

"Not all, my little friend," answered Lucy, "but I must confess that some of them have fallen victims to my neglect: however, there are two fine ones alive, which I shall, with the surviving linnet, cheerfully resign to the care of my dear Harriet, whose tenderness will, I hope, be rewarded by the pleasure of hearing them sing when they are old enough. But I beg you will stay and rest yourselves after your walk."

"Let me see the birds first," said Frederick. "That you shall do," answered Lucy; and taking him by the hand, she conducted him to the room in which she kept them, accompanied by Harriet. Lucy then fed the birds, and gave particular instructions for making their food, and declared that she would never be a receiver of birds' nests any more; but expressed her apprehensions that it would be difficult to wean Edward from his propensity for taking them.

Lucy then took her young friends into the parlour to her governess (for her mamma was dead), who received them very kindly, and gave each of them a piece of cake and some fruit; after which Lucy led them again into the room where the birds were, and very carefully put the nest with the poor solitary linnet into one basket, and that with the two blackbirds into the other. Frederick was very urgent to carry the latter, which his sister consented to; and then bidding adieu to their friend, they set off on their way home, attended by the maid as before.

Let us now return to the redbreasts, whom we left on the wing flying back to the ivy wall, in order to take their young ones abroad.

As the father entered the nest he cried out with a cheerful voice, "Well, my nestlings, are you all ready?" "Yes," they replied. The mother then advanced, and desired that each of them would get upon the edge of the nest. Robin and Pecksy sprang up in an instant, but Dicky and Flapsy, being timorous, were not so expeditious.

The hearts of the parents felt great delight at the view they now had of their young family, which appeared to be strong, vigorous, and lively, and, in a word, endowed with every gift of nature requisite to their success in the world.

"Now," said the father, "stretch your wings, Robin, and flutter them a little in this manner" (showing him the way), "and be sure to observe my directions exactly. Very well," said he: "do not attempt to fly yet, for here is neither air nor space enough for that purpose. Walk gently after me to the wall; then follow me to the tree that stands close to it, and hop on from branch to branch as you will see me do: then rest yourself; and as soon as you see me fly away, spread your wings, and exert all the strength you have to follow me."

Robin acquitted himself to admiration, and alighted very safely on the ground.

"Now stand still," said the father, "till the rest join us." Then going back, he called upon Dicky to do the same as his brother had done; but Dicky was very fearful of fluttering his wings, for he was a little coward, and expressed many apprehensions that he should not reach the ground without falling, as they were such a great height from it. His father, who was a very courageous bird, was quite angry with him.

"Why, you foolish little thing!" said he, "do you mean to stay in the nest by yourself and starve? I shall leave off bringing you food, I assure you. Do you think your wings were given you to be always folded by your sides, and that the whole employment of your life is to dress your feathers and make yourself look pretty? Without exercise you cannot long enjoy health; besides, you will soon have your livelihood to earn, and therefore idleness would in you be the height of folly. Get up this instant."

Dicky, intimidated by his father's displeasure, got up, and advanced as far as the branch from which he was to descend; but here his fears returned, and instead of making an effort to fly, he stood flapping his wings in a most irresolute manner, and suffered his father to lead the way twice without following him. This good parent, finding he would not venture to fly, took a circuit unperceived by Dicky, and watching the opportunity when his wings were a little spread, came suddenly behind him and pushed him off the branch. Dicky, finding himself in actual danger of falling, now gladly stretched his pinions, and upborne by the air, he gently descended to the ground, so near the spot where Robin stood, that the latter easily reached him by hopping.

The mother now undertook to conduct Flapsy and Pecksy, whilst the father stayed to take care of the two already landed. Flapsy made a thousand difficulties, but at length yielded to her mother's persuasions, and flew safely down. Pecksy, without the least hesitation, accompanied her, and by exactly following the directions given, found the task much easier than she expected.

As soon as they had a little recovered from the fatigue and fright of their first essay at flying, they began to look around them with astonishment. Every object on which they turned their eyes excited their curiosity and wonder. They were no longer confined to a little nest built in a small hole, but were now at full liberty in the open air. The orchard itself appeared to them to be a world. For some time each remained silent, gazing round, first at one thing, then at another; at length Flapsy cried out, "What a charming place the world is! I had no conception that it was half so big!"

"And do you suppose then, my dear," replied the mother, "that you now behold the whole of the world? I have seen but a small part of it myself, and yet have flown over so large a space, that what is at present within our view appears to me a little inconsiderable spot; and I have conversed with several foreign birds, who informed me that the countries they came from were so distant that they were many days on their journey hither, though they flew the nearest way, and scarcely allowed themselves any resting-time."

"Come," said the father, "let us proceed to business; we did not leave the nest merely to look about us. You are now, my young ones, safely landed on the ground; let me instruct you what you are to do on it. Every living creature that comes into the world has something allotted him to perform, therefore he should not stand an idle spectator of what others are doing. We small birds have a very easy task, in comparison of many animals I have had an opportunity of observing, being only required to seek food for ourselves, build nests, and provide for our young ones till they are able to procure their own livelihood. We have indeed enemies to dread; hawks and other birds of prey will catch us up if we are not upon our guard; but the worst foes we have are those of the human race, though even among them we redbreasts have a better chance than many other birds, on account of a charitable action which two of our species are said to have performed towards a little boy and girl,* who were lost in a wood, where they were starved to death. The redbreasts saw the affectionate pair, hand in hand, stretched on the cold ground, and would have fed them had they been capable of receiving nourishment; but finding the poor babies quite dead, and being unable to bury them, they resolved to cover them with leaves. This was an arduous task, but many a redbreast has since shared the reward of it; and I believe that those who do good to others always meet with a recompence some way or other. But I declare I am doing the very thing I was reproving you for—chattering away when I should be minding business. Come, hop after me, and we shall soon find something worth having. Fear nothing, for you are now in a place of security; there is no hawk near, and I have never seen any of the human race enter this orchard but the monsters who paid you visits in the nest, and others equally inoffensive."

The father then hopped away, followed by Robin

* Alluding to the ballad of the Children in the Wood.

and Dicky, while his mate conducted the female part of the family. The parents instructed their young ones in what manner to seek for food, and they proved very successful, for there were many insects just at hand.

CHAPTER VIII.

FREDERICK DISCOVERS THE YOUNG ROBINS IN THE CURRANT BUSH.

WHILST all the business related in the last chapter was going on in the redbreast family, Harriet and her brother were walking home with the poor birds in the baskets. "Well, Frederick," said she to him, "what think you of bird-nesting now? Should you like to occasion the deaths of a number of little harmless creatures?"

"No, indeed," said Frederick; "and I think Miss Jenkins a very naughty girl for starving them."

"She was to blame, but is now sorry for her fault, my dear, therefore you must not speak unkindly of her; besides, you know she has no good mamma, as we have, to teach her what is proper; and her papa is obliged to be absent from home very often, and leave her to the care of a governess, who perhaps was never instructed herself to be tender to animals."

With this kind of conversation they amused themselves as they walked, every now and then peeping into their baskets to see the little birds, which were very lively and well. They entreated the maid to take them through the orchard, which had a gate that opened into a meadow that lay in their way, having no doubt of obtaining admittance, as it was the usual hour for their friend Joe to work there. They accordingly knocked at the gate, which was immediately opened to them, and Frederick requested Joe to show him the robins' nest.

Just at this time the young robins were collected together near the gate, when they were suddenly alarmed with a repetition of the same noises which had formerly terrified them in the nest; and Robin, who was foremost, beheld, to his very great amazement, Frederick and Harriet, the maid who attended them, with Joe the gardener, who, having opened the gate, was, at the request of his young master and mistress, conducting them to the ivy wall.

Robin, with all his courage (and, indeed, he was not deficient in this quality), was seized with a great tremor; for if the view he had of the faces of these persons had appeared so dreadful to him when he sat in the nest, what must it now be to behold their full size, and see them advancing with, as he thought, gigantic strides towards him? He expected nothing less than to be crushed to death with the foot of one of them; and not having yet attained his full strength, and never having raised himself in the air, he knew not how to escape, therefore chirped so loudly as not only to surprise his brother and sisters, and bring his father and mother to inquire the meaning of his cry, but also to attract the attention of the young Bensons.

'What chirping is that?" cried Harriet.

"It was the cry of a young bird," said the maid; " was it not one of those in the baskets?"

"No," said Frederick, "the noise came that way," pointing to some currant bushes; "my birds are very well."

"And so is my linnet," replied Harriet.

Frederick then set down his charge very carefully and began looking about in the place from whence he supposed the sound proceeded, when, to his great joy, he soon discovered the redbreasts and their little family. He called eagerly to his sister, who was equally pleased with the sight. They then stooped down to take a nearer view of them, by which means he directly confronted Robin, who, as soon as the

G

young gentleman's face was on a level with his eyes, recollected him, and calling to his brother and sisters, told them they need not be afraid.

Harriet followed her brother's example, and delighted the little flock with the sight of her amiable countenance. She heartily lamented having nothing with which to regale her old favourites and their family, when Frederick produced from his pocket a piece of biscuit, which they crumbled and scattered. Harriet, recollecting that her mamma would expect her at home, and that the birds in the baskets would be hungry, persuaded her brother to take up his little load and return. They therefore left the redbreasts enjoying the fruits of their bounty.

When the happy birds had shared amongst them the kind present of their young benefactors, they hopped about in search of some moister food. Dicky had the good fortune to find four little worms together, but instead of calling his brother and sisters to partake of them, he devoured them all himself.

"Are you not ashamed, you little greedy creature?" cried his father, who observed his selfish disposition. "What would you think of your brother and sisters were they to serve you so? In a family every individual ought to consult the welfare of the whole, instead of his own private satisfaction;

it is his own truest interest to do so. A day may come when he who has now sufficient to supply the wants of his relations may stand in need of assistance from them. But setting aside selfish considerations, which are the last that ever find place in a generous breast, how great is the pleasure of doing good, and contributing to the happiness of others!"

Dicky was quite confounded, and immediately hopped away to find, if possible, something for his brother and sisters, that he might regain their good opinion.

In the meanwhile Robin found a caterpillar, which he intended to take for Pecksy; but just as he was going to pick it up, a linnet, which had a nest in the orchard, snatched it from him, and flew away with it.

With the most furious rage Robin advanced to his father, and entreated that he would fly after the linnet and tear his heart out.

"That would be taking violent revenge indeed," said his father. "No, Robin, the linnet has as great a right to the caterpillar as you or I, and in all probability he has as many little gaping mouths at home ready to receive it. But however this may be, I had for my own part rather sustain an injury than take revenge. You must expect to have many a scramble of this kind in your life; but if you give way to a resentful temper, you will do yourself more harm than all the enemies in the world can do you, for you will be in perpetual agitation, from an idea that every one who does not act in direct conformity with your wishes has a design against you. Therefore restrain your anger, that you may be happy; for, believe me, peace and tranquillity are the most valuable things you can possess."

At this instant Pecksy came up with a fine fat spider in her mouth, which she laid down at her mother's feet, and thus addressed her:—

"Accept, my dear parent, the first tribute of gratitude which I have ever been able to offer you. How have I formerly longed to ease those toils which you and my dear father have endured for our sakes! and gladly would I now release you from further fatigue on my account; but I am still a poor creature, and must continue to take shelter under your wing. I will hop, however, as long as I am able, to procure food for the family."

The eyes of the mother sparkled with delight, and knowing that Pecksy's love would be disappointed by a refusal, she ate the spider which the dutiful nestling had so affectionately brought her, and then said,—

"How happy would families be if every one, like you, my dear Pecksy, consulted the welfare of the rest, instead of turning their whole attention to their own interest!"

Dicky was not present at this speech, which he might have considered as a reflection on his own conduct; but he arrived as it was ended, and presented Pecksy with a worm, like those he had himself so greedily eaten. She received it with thanks, and declared it was doubly welcome from his beak.

"Certainly," said the mother, "fraternal love stamps a value on the most trifling presents."

Dicky felt himself happy in having regained the good opinion of his mother and obliged his sister, and resolved to be generous for the future.

The mother bird now reminded her mate that it would be proper to think of returning to the nest.

"If the little ones fatigue themselves too much with hopping about," said she, "their strength will be exhausted, and they will not be able to fly back."

"True, my love," replied her mate; "gather them under your wings a little, as there is no reason to apprehend danger here, and then we will see what they can do."

She complied with his desire, and when they were sufficiently rested she got up, on which the whole brood instantly raised themselves on their feet.

"Now, Robin," cried the father, "let us see your dexterity in flying upwards: come, I will show you how to raise yourself."

"Oh, you need not take that trouble," said the conceited bird; "as I flew down, I warrant I know how to fly up."

Then spreading his wings, he attempted to rise, but in so unskilful a manner that he only shuffled along upon the ground.

"That will not do, however," cried the father; "shall I show you now?"

Robin persisted in it that he stood in no need of instruction, and tried again: he managed to raise himself a little way, but soon tumbled headlong. His mother then began reproving him for his obstinacy, and advised him to accept his father's kind offer of teaching him.

"You may depend on it, Robin," said she, "that he is in every respect wiser than you, and as he has had so much practice, he must of course be expert in the art of flying; and if you persist in making your own foolish experiments, you will only commit a number of errors, and make yourself ridiculous. I should commend your courage, provided you would add prudence to it; but blundering on in this ignorant manner is only rashness."

"Let him alone, let him alone," said the father; "if he is above being taught, he may find his own way to the nest; I will teach his brother.—Come, Dicky," said he, "let us see what you can do at flying upwards; you cut a noble figure this morning when you flew down."

Dicky, with reluctance, advanced; he said he did not see what occasion they had to go back to the nest at all; he should suppose they might easily find some snug corner to creep into till they were strong enough to roost in trees, as other birds did.

"Why," said the father, "you are as ridiculous with your timidity as Robin with his conceit. Those who give way to groundless fears generally expose themselves to real dangers. If you rest on the earth all night, you will suffer a great deal from cold and damp, and may very likely be devoured whilst you sleep, by rats and other creatures that go out in the night to seek for food; whereas, if you determine to go back to the nest, you have but one effort to make, for which, I will venture to say, you have sufficient strength, and then you will lie warm, safe, and quiet: however, do as you will."

Dicky began to think that it was his interest to obey his father, and said he would endeavour to fly up, but was still fearful he should not be able to do it.

"Never despair," replied his father, "of doing what others have done before you. Turn your eyes upwards, and behold what numbers of birds are at this instant soaring in the air. They were once all nestlings, like yourself. See there that new-fledged wren, with what courage he skims along Let it not be said that a redbreast lies grovelling on the earth while a wren soars above him!"

Dicky was now ashamed of himself, and inspired with emulation, therefore without delay he spread his wings and his tail; his father with pleasure placed himself in a proper attitude before him, then rising from the ground, led the way; and Dicky, by carefully following his example, safely arrived at the nest, which he found a most comfortable resting-place after the fatigue of the morning, and rejoiced that he had a good father to teach him what was most conducive to his welfare.

The father, having seen him safe home, returned to his mate, who, during his short absence, had been endeavouring to convince Robin of his fault, but to no purpose; he did not like to be taught what he still persuaded himself he could do by his own exertions; she therefore applied herself to Flapsy.

"Come, my dear," said she, "get ready to follow me when your father returns, for the sun casts a great heat here, and the nest will be quite comfortable to you." Flapsy dreaded the experiment; however, as she could not but blame both Robin and Dicky's conduct, she resolved to do her best, but entreated her mother to inform her very particularly how to proceed. "Well, then," said the tender parent, "observe me. First bend your legs then spring from the ground as quick as you can, stretching your wings straight out on each side of your body as you rise; shake them with a quick motion, as you will see me do, and the air will yield to you, and at the same time support your weight; whichever way you want to turn, strike the air with the wing on the contrary side, and that will bring you about." She then rose from the ground, and having practised two or three times repeatedly what she had been teaching, Flapsy at length ventured to follow her, but with a palpitating heart; and was soon happily seated in the nest by the side of

Dicky, who rejoiced that his favourite sister was safely arrived.

The mother bird now went back to Pecksy, who was waiting with her father till she returned; for the good parent chose to leave the female part of his family to the particular management of their mother.

Pecksy was fully prepared for her flight, for she had attentively observed the instruction given to the others, and also their errors; she therefore kept the happy medium betwixt self-conceit and timidity, indulging that moderated emulation which ought to possess every young heart; and resolving that neither her inferiors nor equals should soar above her, she sprang from the ground, and, with a steadiness and agility wonderful for her first essay, followed her mother to the nest, who, instead of stopping to rest herself there, flew to a neighbouring tree, that she might be at hand to assist Robin, should he repent of his folly. But Robin disappointed her hopes, for he sat sulky; though convinced he had been in the wrong, he would not humble himself to his father, who therefore resolved to leave him a little while, and return to the nest.

As soon as Robin found himself deserted, instead

of being sorry, he gave way to anger and resentment. "Why," cried he, "am I to be treated in this manner, who am the eldest in the family, while all the little darlings are fondled and caressed? But I don't care; I can get to the nest yet, I make no doubt." He then attempted to fly, and after a great many trials at length got up in the air; but not knowing which way to direct his course, he sometimes turned to the right and sometimes to the left now he advanced forwards a little, and now, fearing he was wrong, came back again; at length, quite spent with fatigue, he fell to the ground, and bruised himself a good deal: stunned with the fall, he lay for some minutes without sense or motion, but soon revived; and finding himself alone in this dismal condition, the horrors of his situation filled him with dreadful apprehensions and the bitterest remorse.

"Oh," cried he, "that I had but followed the advice and example of my tender parents! then had I been safe in the nest, blest with their kind caresses, and enjoying the company of my dear brother and sisters; but now I am of all birds the most wretched! Never shall I be able to fly, for every joint of me has received a shock which I doubt it will not recover. Where shall I find shelter from the scorching sun, whose piercing rays already render the ground I lie on intolerably hot? What kind beak will supply me with food to assuage the pangs of hunger which I shall soon feel? By what means shall I procure even a drop of water to quench that thirst which so frequently returns? Who will protect me from the various tribes of barbarous animals which I have been told make a prey of birds? Oh, my dear, my tender mother! if the sound of my voice can reach your ears, pity my condition, and fly to my succour!"

The kind parent waited not for further solicitation, but parting from the branch on which she had been a painful eye-witness of Robin's fall, she instantly stood before him.

"I have listened," said she, "to your lamentations, and since you seem convinced of your error, I will not add to your sufferings by my reproaches; my heart relents towards you, and gladly would I afford you all the aid in my power; but, alas! I can do but little for your relief. However, let me persuade you to exert all the strength you have, and use every effort for your own preservation; I will endeavour to procure you some refreshment, and at the same time contrive the means of fixing you in a place of more security and comfort than that in which you at present lie." So saying, she flew to a little stream which flowed in an adjacent meadow, and fetched from the brink of it a worm, which she had observed an angler to drop as she perched on the tree; with this she immediately returned to the penitent Robin, who received the welcome gift with gratitude.

Refreshed with this delicious morsel, and comforted by his mother's kindness, he was able to stand up, and on shaking his wings, he found that he was not so greatly hurt as he apprehended; his head, indeed, was bruised, so that one eye was almost closed, and he had injured the joint of one wing, so that he could not possibly fly: however, he could manage to hop, and the parent bird observing that Joe the gardener was cutting a hawthorn hedge which was near the spot, desired Robin to follow her. This he did, though with great pain. "Now," said she, "look carefully about, and you will soon find insects of one kind or another for your sustenance during the remainder of the day, and before evening I will return to you again. Summon all your courage, for I make no doubt you will be safe while our friend continues his work, as none of those creatures which are enemies to birds will venture to come near him." Robin took a sorrowful farewell, and the mother flew to the nest.

CHAPTER IX.

THE VISIT TO MRS. ADDIS'S.

"You have been absent a long time, my love," said her mate; "but I perceived that you were indulging your tenderness towards that disobedient nestling, who has rendered himself unworthy of it. However, I do not condemn you for giving him assistance, for had not you undertaken the task, I would myself have flown to him instead of returning home. How is he?—likely to live and reward your kindness?" "Yes," said she, "he will, I flatter myself, soon perfectly recover, for his hurt is not very considerable; and I have the pleasure to tell you he is extremely sensible of his late folly, and I dare say will endeavour to repair his fault with future good behaviour."

"This is pleasing news indeed!" said he.

The little nestlings, delighted to hear their dear brother was safe and convinced of his error, expressed great joy and satisfaction, and entreated their father to let them descend again and keep him company. To this he would by no means consent, because, as he told them, the fatigue would be too great, and it was proper that Robin should feel a little longer the consequences of his presumption. "To-morrow," said he, "you shall pay him a visit, but to-day he shall be by himself." On this they dropped their request, knowing that their parent was the best judge of what was proper to be done, and not doubting but that his affection would lead him to do everything that was conducive to the real happiness of his family; but yet they could not tell how to be happy without Robin, and were continually perking up their little heads, fancying they heard his cries. Both their father and mother frequently took a peep at him, and had the satisfaction of seeing him very safe by their friend Joe the gardener, though the honest fellow did not know of his own guardianship, and continued his work without perceiving the little cripple, who hopped and shuffled about, pecking here and there whatever he could meet.

When he had been for some time by himself, his

mother made him another visit, and told him she had interceded with his father, whose anger was abated, and he would come to him before he went to rest. Robin rejoiced to hear that there was a chance of his being reconciled to his father, yet he dreaded the first interview; however, as it must be, he wished to have it over as soon as possible, and every wing he heard beat the air he fancied to be that of his offended parent. In this state of anxious expectation he continued almost to the time of sunset, when of a sudden he heard the well-known voice to which he used to listen with joy, but which now caused his whole frame to tremble; but observing a beam of benignity in that eye in which he looked for anger and reproach, he cast himself in the most supplicating posture at the feet of his father, who could no longer resist the desire he felt to receive him into favour.

"Your present humility, Robin," said he, "disarms my resentment; I gladly pronounce your pardon, and am persuaded you will never again incur my displeasure. We will therefore say no more on a subject which gives so much pain to us."

"Yes, my dear indulgent father," cried Robin, "permit me to make my grateful acknowledgments for your kindness, and to assure you of my future obedience." The delighted parent accepted his submission, and the reconciliation was completed.

By this time Robin was greatly exhausted; his kind father therefore conducted him to a pump in the garden, where he refreshed himself with a few drops of water. He now felt himself greatly relieved; but on his father's asking him what he intended to do with himself at night, his spirit sank again, and he answered, he did not know.

"Well," said the father, "I have thought of an expedient to secure you from cold at least. In a part of the orchard, a very little way from here, there is a place belonging to our friend the gardener, where I have sheltered myself from several storms, and am sure it will afford you a comfortable lodging; so follow me before it is too late."

The old bird then led the way, and his son followed him. When they arrived, they found the door of the tool-house open, and as the threshold was low, Robin managed to get over it. His father looked carefully about, and at last found in a corner a parcel of shreds, kept for the purpose of nailing up trees. "Here, Robin," said he, "is a charming bed for you; let me see you in it and call your mother to have a peep, and then I must bid you " Good night."

So saying, away he flew, and brought his mate, who was perfectly satisfied with the lodging provided for her late undutiful but now repentant son; but, reminded by her mate that if they stayed longer they might be shut in, they took leave, telling Robin they would visit him early in the morning.

Though this habitation was much better than Robin expected, and he was ready enough to own better than he deserved, yet he deeply regretted his absence from the nest, and longed to see again his brother and sisters. However, though part of the night was spent in bitter reflections, fatigue at length prevailed over anxiety, and he fell asleep. The nestlings were greatly pleased to find that Robin was likely to escape the dangers of the night, and even the anxious mother at length resigned herself to repose.

Before the sun showed his glorious face in the east, every individual of this affectionate family was awake: the father with impatience waited for the gardener's opening the tool-house; the mother prepared her little ones for a new excursion.

"You will be able to descend with more ease, my dears, to-day than you did yesterday, shall you not?"

"Oh yes, mother," said Dicky, "I shall not be at all afraid." "Nor I," said Flapsy.

"Say you so? then let us see which of you will be down first," said she. "Come, I will show you the way."

On this, with gradual flight the mother bent her course to a spot near the place where Robin lay concealed; they all instantly followed her, and surprised their father, who having seen Joe, was every instant expecting he would open the door. At length, to the joy of the whole party, the gardener appeared, and they soon saw him fetch his shears and leave the tool-house open; on this the mother proposed that they should all go together and call Robin. There they found him in his snug little bed: but who can describe the happy meeting? who can find words to express the raptures which filled each little bosom?

When the first transports subsided, "I think," said the father, "it will be best to retire from hence. If our friend returns, he may take us for a set of thieves, and suppose that we came to eat his seeds, and I should be sorry he should have an ill opinion of us." "Well, I am ready," said his mate. "And we!" cried the whole brood.

They accordingly left the tool-house, and hopped about among the currant bushes. "I think," said the father, "that you who have the full use of your limbs could manage to get up these low trees, but Robin must content himself upon the ground a little longer." This was very mortifying, but he had no one to blame excepting himself; so he forbore to complain, and assumed as much cheerfulness as he could. His brothers and sisters begged they might stay with him all day, as they could do very well without going up to the nest; to this the parents consented.

It is now time to inquire after Harriet and her brother. These happy children reached home soon after they left the redbreasts, and related every circumstance of their expedition to their kind mamma, who, hearing the little prisoners in the basket chirp very loudly, desired they would immediately go and feed them, which they gladly did, and then took a short lesson. Mrs. Benson told Harriet that she was going to make a visit in the afternoon, and should take her with her, therefore desired she would keep herself quite still, that she might not be fatigued after the walk she had had in the morning; for though she meant to go in the carriage it was her intention to walk home, as the weather was so remarkably fine. The young lady took great care of the birds, and Frederick engaged, with the assistance of the maid, to feed them during her absence. Harriet was then dressed to attend her mamma.

Mrs. Addis, to whose house they were going, was a widow lady; she had two children,—Charles, a boy of twelve years old, at school, and Augusta, about seven, at home. But these children were quite strangers to the Bensons.

On entering the hall Harriet took notice of a very disagreeable smell, and was surprised with the appearance of a parrot, paroquet, and a macaw, all in most superb cages. In the next room she came to were a squirrel and a monkey, which had each a little house neatly ornamented. On being introduced into the drawing-room she observed in a corner a lap dog lying on a splendid cushion; and in a beautiful little cradle, which she supposed to contain a large wax doll, lay in great state a cat with a litter of kittens.

After the usual compliments were over, Mrs. Benson said, " I have taken the liberty of bringing my daughter with me, in hopes of inducing you to favour us, in return, with the company of Miss Addis."

"You are very obliging," replied the lady, "but indeed I never take my children with me, they are so rude; it will be time enough some years hence for Augusta to go visiting."

"I am sorry to hear you say this," said Mrs. Benson. "You are displeased, then, I fear, at my having brought Harriet with me." This in reality was the case, as Mrs. Benson plainly perceived, for the lady made no answer, and looked very cross.

Harriet was curious to examine the variety of animals which Mrs. Addis had collected together; but as her mamma never suffered her to run about when she accompanied her to other people's houses, she sat down, only glancing her eye first to one part of the room and then to the other, as her attention was successively attracted.

As Mrs. Benson requested to see Miss Addis, her mamma could not refuse sending for her; she therefore rang the bell, and ordered that Augusta might come down to her. The footman, who had never before received such a command (for Mrs. Addis only saw the child in the nursery), stared with astonishment, and thought he had misunderstood it. However, on his lady repeating her words, he went up-stairs to tell the nursery-maid the child was to be taken to the drawing-room. "What new fancy is this?" said she; "who would ever have thought of her wanting the child in the drawing-room? I have no stockings clean for her, nor a frock to put on but what is ragged. I wish she would spend less money on her cats and dogs and monkeys, and then her child would appear as she ought to do." "I won't go down-stairs, Nanny," said the child. "But you must," said Nanny; "besides, there's a pretty young lady come to see you, and if you go like a good girl, you shall have a piece of sugared bread and butter for your supper; and you shall carry the new doll which your godmamma gave you, to show to your little visitor."

These bribes had the desired effect, and Miss Addis went into the drawing-room; but instead of entering it like a young lady, she stopped at the door, hung down her head, and looked like a little simpleton. Harriet was so surprised at her awkwardness that she did not know what to do, and looked at her mamma, who said, "Harriet, my love, can't you take the little girl by the hand and lead her to me? I believe she is afraid of strangers." On this Harriet rose to do so; but Augusta, apprehensive that she would snatch her doll away, was going to run out, only she could not open the door.

Mrs. Benson was quite shocked to see how sickly, dirty, and ragged this poor child was, and how vulgar also, for want of education; but Mrs. Addis was so taken up at that instant with the old lap dog, which had, as she thought, fallen into a fit, that she did not notice her entrance; and before she perceived it, the child went up to the cradle in order to put her doll into it, and seized one of the kittens by the neck, the squeaking of which provoked the old cat to scratch her, and this made her cry and drop the kitten upon the floor. Mrs. Addis, seeing this, flew to the little animal, endeavoured to soothe it with caresses, and was going to beat Augusta for touching it, but Mrs. Benson interceded for her; she was, however, sent away into the nursery. Happily for children, there are not many such mammas as Mrs. Addis.

The tea-things being set, the footman came in with the urn; and, both his hands being employed, he left the door open. To the great terror of Harriet, and even of her mamma too, he was followed by the monkey they saw in the hall, which, having broken his chain, came to make a visit to his lady. Mrs. Addis, far from being disconcerted, seemed highly pleased with his cleverness. "Oh, my sweet dear Pug!" said she, "are you come to see us? Pray show how like a gentleman you can behave." Just as she had said this he leaped upon the tea-table, and took cup after cup and threw them on the ground till he broke half the set; then jumped on the sofa and tore the cover of it; in short, as soon as he had finished one piece of mischief he began another, till Mrs. Addis, though greatly diverted with his wit, was obliged to have him caught and confined; after which she began making tea, and quietness was for a short time restored. But Mrs. Benson, though capable of conversing on most subjects, could not engage Mrs. Addis in any discourse but upon the perfection of her birds and beasts, and a variety of uninteresting particulars were related concerning their wit or misfortunes.

On hearing the clock strike seven, Mrs. Addis begged Mrs. Benson to excuse her, as she made it a constant rule to see all her dear darlings fed at that hour, and entreated that she and the young lady would take a turn in the garden in the meanwhile. This was very ill-bred, but Mrs. Benson desired she would use no ceremonies with her, and was really glad of the respite it gave her from company so irksome, and Harriet was happy to be alone with her mamma; she, however, forbore making any remarks on Mrs. Addis, because she had been taught that it did not become young persons to censure the behaviour of those who were older than themselves.

The garden was spacious, but overrun with weeds; the gravel walks were so rough for want of rolling that it was quite painful to tread on them, and the grass on the lawn so long that there was no walking with any comfort, for the gardener was almost continually going on some errand or another for Mrs. Addis's darlings; so Mrs. Benson and her daughter sat down on a garden seat, with an intention of waiting there till Mrs. Addis should summon them.

Harriet could not refrain from expressing a wish that it was time to go home; to which Mrs. Benson replied that she did not wonder at her desire to return; "But," said she, "my dear, as the world was not made merely for us, we must endeavour to be patient under every disagreeable circumstance we meet with. I know what opinion you have formed of Mrs. Addis, and should not have brought you to be a spectator of her follies, had I not hoped that an hour or two passed in her company would afford you a lesson which might be useful to you through life. I have before told you that our affections towards the inferior parts of the creation should be properly regulated; you have, in your friend Lucy Jenkins and her brother, seen instances of cruelty to them which I am sure you will never be inclined to imitate; but I was apprehensive you might fall into the contrary extreme, which is equally blameable. Mrs. Addis, you see, has absolutely transferred the affection which she ought to feel for her child to creatures which would really be much happier without it. As for Puss, who lies in the cradle in all her splendour, I will engage to say she would pass her time pleasanter in a basket of clean straw, placed in a situation where she could occasionally amuse herself with catching mice. The lapdog is, I am sure, a miserable object, full of diseases, the consequence of luxurious living. How enviable is the lot of a spaniel that is at liberty to be the companion of his master's walks, when compared with his! Pug, I am certain, would enjoy himself much more in his native woods; and I am greatly mistaken if the parrots, &c., have not cause to wish themselves in their respective countries, or at least divided into separate families, where they would be better attended; for Mrs. Addis, by having such a number of creatures, has put it out of her power to see properly with her own eyes to all. But come, let us go back into the house; the time for our going home draws near, and I do not wish to prolong my visit."

Saying this, Mrs. Benson arose, and with her daughter went into the drawing-room, which opened into the garden; the other door, which led to the adjoining apartments, was not shut; this gave them an opportunity of hearing the following discourse, which greatly distressed Mrs. Benson, and perfectly terrified the gentle Harriet.

"Begone, wretch!" said Mrs. Addis, "begone this instant! you shall not stay a moment longer in this house!" "I hope, madam, you will have the goodness to give me a character; indeed and indeed, I fed Poll, but I believe he got cold when you let him stand out of doors the other day." "I will give you no character, I tell you," said Mrs. Addis, "so depart this instant. Oh, my poor dear, dear creature! I fear you will never recover.—John!—Thomas! here, run this instant to Perkins, the birdcatcher; perhaps he can tell me what to give him." Then bursting into a flood of tears, she sat down and forgot her guests.

Mrs. Benson thought it necessary to remind her that she was in the house, and stepped to the door to ask what was the matter: Mrs. Addis recollected herself sufficiently to beg pardon for neglecting to pay attention to her, but declared that the dreadful misfortune that had befallen her had made her insensible to everything else.

"What can be the matter?" said Mrs. Benson, "have you heard of the death of a dear friend? has your child met with an accident?" "Oh no," said she, "but poor Poll is taken suddenly ill—my dear Poll, which I have had these seven years,—and I fear he will never recover."

"If this is all, madam," said Mrs. Benson, "I really cannot pity you, nor excuse your behaviour to me, for it is an instance of disrespect which I believe no other person but yourself would show me, and I shall take my leave of your house for ever; but before I go, permit me to say that you act in a very wrong manner, and will certainly feel the ill effects of your injustice to your fellow-creatures, in thus robbing them of the love you owe them, to lavish it away on those animals, which are really sufferers by your kindness."

At this instant the footman entered to inform Mrs. Benson that her servant was come; on which, accompanied by her daughter, she without further ceremony left Mrs. Addis to compose herself as she could.

CHAPTER X.

ADVENTURES OF THE LITTLE ROBINS.

As they walked along, both Mrs. Benson and her daughter continued silent, for the former was greatly agitated, and the latter quite in consternation at what had lately passed. But their attention was soon awakened by the supplication of a poor woman, who entreated them to give her some relief, as she had a sick husband and seven children in a starving condition; of which, she said, they might be eye-witnesses if they would have the goodness to step into a barn that was very near. Mrs. Benson, who was always ready to relieve the distressed, taking her daughter by the hand, and desiring the servant to stop for her, followed the woman, who conducted her to the abode of real woe, where she beheld a father surrounded by his helpless family, whom he could no longer maintain, and who, though his disease was nearly subdued, was himself almost ready to die for want of good nourishing diet.

"How came you all to be in this condition, good woman?" said Mrs. Benson to his wife; "surely you might have obtained relief before your husband was reduced to such extremity?" "Oh, my good lady," said the woman, "we have not been used to begging, but to earn an honest livelihood by our industry; and never till this sad day have I known what it was to ask charity. This morning, for the first time, I applied at the only great house in this village, where I made no doubt there was abundance. I told my dismal tale to a servant, and begged she would make it known to her mistress; but she assured me it was in vain to come there, for her lady had such a family of cats, dogs, monkeys, and all manner of creatures, that she had nothing to spare for poor people; at the same instant I saw the poulterer bring a rabbit and a fowl, which I found were for the favourite cat and dog. This discouraged me from begging; and I had determined that I never would ask again, but the sight of my dear husband and children in this condition drove me to do it."

"Well, comfort yourself," said Mrs. Benson, "we will see what we can do; in the meantime here is something for a present supply." Mrs. Benson then departed, as she was fearful of being late. Harriet was greatly affected at this scene, and could no longer help exclaiming against Mrs. Addis.

"She is deserving of great blame, indeed," said Mrs. Benson; "but I have the pleasure to say, such characters as hers are very uncommon—I mean in the extreme, though there are numbers of people who fall into the same fault in some degree, and make themselves truly ridiculous with their unnatural affections. I wish you, while you are young, to guard your mind against such a blameable weakness."

Harriet assured her mamma that she should never forget either Mrs. Addis or the lesson she had received on the subject, and then expressed her satisfaction that they had met the poor woman. "I rejoice sincerely," said Mrs. Benson, "at having been fortunate enough to come in time to assist this poor miserable family, and hope, my love, you will, out of your own little purse, contribute something towards their relief." "Most willingly," said Harriet; "they shall be welcome to my whole store."

They kept talking on this subject till they arrived at home. Little Frederick, who sat up an hour beyond his time, came out to meet them, and assured his sister that the birds were well, and fast asleep. "I think," said she, "it is time for you and me to follow their example; for my part, with my morning and evening walk together, I am really tired, so shall beg leave to wish you a good night, my dear mamma. Papa, I suppose, will not be at home this week?" "No, my dear, nor the next," said Mrs. Benson, "for he has many affairs to settle in the west. I am rather fatigued also, and shall soon retire to rest."

At the usual hour of visiting Mrs. Benson's tea-table the next day, the parent robins took their morning's flight, and found the children with their mother. They had been up a long time, for Frederick had made in his bedchamber a lodging for the birds, which had awakened both him and his sister at a very early hour, and they rose with great readiness to perform the kind office they had imposed upon themselves.

The two blackbirds were perfectly well, but the linnet looked rather drooping, and they began to be apprehensive they should not raise him, especially when they found he was not inclined to eat. As for the blackbirds, they were very hungry indeed; and their young benefactors, not considering that when fed by their parents young birds wait some time between every morsel, supplied them too fast, and filled their crops so full that they looked as if they had great wens on their necks; and Harriet perceived one of them gasping for breath.

"Stop, Frederick!" said she, as he was carrying the quill to its mouth; "the bird is so full he can hold no more." But she spoke too late; the little creature gave his eyes a ghastly roll, and fell on one side, suffocated with abundance.

"Oh, he is dead! he is dead!" cried Frederick.

"He is indeed," said Harriet; "but I am sure we did not mean to kill him, and it is some satisfaction to think that we did not take the nest."

This consideration was not enough to comfort Frederick, who began to cry most bitterly; his mamma hearing him, was apprehensive he had hurt himself, for he seldom cried unless he was in great pain; she therefore hastily entered the room to inquire what was the matter, on which Harriet related the disaster that had happened. Mrs. Benson then sat down, and taking Frederick in her lap, wiped his eyes, and giving him a kiss, said,—

"I am sorry, my love, for your disappointment; but do not afflict yourself; the poor little thing is out of his pain now, and I fancy suffered but for a short time. If you keep on crying so, you will forget to feed your flock of birds, which I fancy, by the chirping I heard from my window, are beginning to assemble. Come, let me take the object of your distress out of your sight; it must be buried." Then, carrying the dead bird in one hand, and leading Frederick with the other, she went down-stairs.

While she was speaking, Harriet had been watching the other blackbird, which she had soon the pleasure to see perfectly at his ease. She then attempted to feed the linnet, but he would not eat.

"I fancy, Miss Benson," said the maid, "he wants air."

"That may be the case, indeed," replied Harriet, "for you know, Betty, this room, which has been shut up all night, must be much closer than the places birds build in." Saying this, she opened the window, and placed the linnet near it, waiting to see the effect of the experiment, which answered her wishes; and she was delighted to behold how the little creature gradually smoothed his feathers, and his eyes resumed their native lustre; she once more offered him food, which he took, and quite recovered. Having done all in her power for her little orphans, she went to share with her brother the task of feeding the daily pensioners, which being ended, she seated herself at the breakfast-table by her mamma.

"I wonder," said Frederick, who had dried up his tears, "that the robins are not come."

"Consider," said his sister, "that they have a great deal of business to do now that their young ones begin to leave their nest; they will be here by and by, I make no doubt." An instant after, they entered the room. The sight of them perfectly restored Frederick's cheerfulness; and after they were departed, he requested of his mamma that he and Harriet might go again into the orchard, in hopes of seeing the young robins.

"That you shall do, Frederick," said she, "upon condition that you continue a very good boy; but as yesterday was rather an idle day with you, you must apply a little closer to-day; and Harriet has a great deal of business to do, therefore you must wait till evening, and then perhaps I may go with you."

Frederick was satisfied with this promise, and took great pains to read and spell. He repeated by heart one of Mrs. Barbauld's hymns, and some other little things which he had been taught; and Harriet applied herself to a variety of different lessons with great diligence, and performed her task of work entirely to her mamma's satisfaction.

As soon as the old redbreasts left their little family in order to go to Mrs. Benson's, Pecksy, with great kindness, began to ask Robin where he had hurt himself, and how he did it.

"Oh," said he, "I am much better; but it is a wonder I am now alive, for you cannot think what a dreadful fall I had. With turning about as I did in the air I became quite giddy, so could not make the least exertion for saving myself as I was falling, and came with great force to the ground; you see how my eye is still swelled, and it was much more so at first. My wing is the worst, and still gives me a great deal of pain; observe how it drags on the ground; but as it is not broke, my father says it will soon be well, and I hope it will be so, for I long to be flying, and shall be glad to receive any instructions for the future. I cannot think how I could be so foolishly conceited as to suppose I knew how to conduct myself without my father's guidance."

"Young creatures like us," said Pecksy, "certainly stand in need of instruction, and we ought to think ourselves happy in having parents who are willing to take the trouble of teaching us what is necessary for us to know. I dread the day when I must quit the nest and take care of myself." Flapsy said she made no doubt they would know how to fly and peck and do everything before that time; and for her part she longed to see the world, and to know how the higher ranks of birds behaved themselves, and what pleasures they enjoyed; and Dicky declared he felt the same wish, though he must confess he had great dread of birds of prey.

"Oh," said Flapsy, "they will never seize such a pretty creature as you, Dicky, I am sure."

"Why, if beauty can prevail against cruelty, you will be secure, my sweet sister," replied he, "for your delicate shape must plead in your behalf."

Just as he had finished his speech a hawk appeared in sight, on which the whole party was seized with a most uncommon sensation, and threw themselves on their backs, screaming with all their might; and at the same instant the cries of numbers of little birds echoed through the orchard. The redbreasts soon recovered, and rising on their feet, looked about to see what was become of the cause of their consternation; when they beheld him high in the air, bearing off some unhappy victim, a few of whose feathers fell near the young family, who, on examining them, found they belonged to a goldfinch; on which Pecksy observed that it was evident these savages paid no attention to personal beauty. Dicky was so terrified, he knew not what to do, and had thoughts of flying back to the nest, but after

Robin's misfortune he was fearful of offending his father; he therefore got up into a currant bush, and hid himself in the thickest part of the leaves. Flapsy followed him; but Robin being obliged to keep on the ground, Pecksy kindly resolved to bear him company.

In a few minutes their parents returned from Mrs. Benson's and found the two latter pretty near where they had left them; but missing the others, the mother with great anxiety inquired what was become of them. Robin then related how they had been frightened by a hawk; and while he was doing so, they returned to him again.

"I am surprised," said the father, "that a hawk should have ventured so near the spot where the gardener was at work." Pecksy informed him that they had not seen the gardener since he left them. "Then I dare say he is gone to breakfast," replied the mother; and this was the case, for they at this instant saw him return with his shears in his hand, and soon pursue his work. "Now you will be safe," cried the father; "I shall therefore stay and teach you to fly in different directions, and then your mother and I will make some little excursions, and leave you to practise by yourselves; but first of all let me show you where to get water, for I am sure you must be very thirsty." "No," said they, "we have had several wet worms and juicy caterpillars, which have served us both for victuals and drink; Robin is very quick at finding them." "There is nothing like necessity to teach birds how to live," said the father; "I am glad Robin's misfortunes have been so beneficial to him." "What would have become of you, Robin, if you had not exerted yourself as I directed?" said his mother; "you would soon have died had you continued to lie on the scorching ground. Remember from this instance, as long as you live, that it is better to use means for your own relief than to spend time in fruitless lamentations. But come along, Dicky, Flapsy, and Pecksy, there is water near." She then conducted them to the pump from which Joe watered the garden, which was near the tool-house where Robin slept.

Here they stayed some time, and were greatly amused, still so near the gardener that they regarded themselves as under his protection. The parents flew up into a tree, and there the father entertained his beloved mate and family with his cheerful music; and sometimes they made various airy excursions for examples to their little ones. In this manner the day passed happily away, and early in the evening Flapsy, Pecksy, and Dicky were conducted to the nest; they mounted in the air with much more ease than the preceding day, and the parents instructed them how to fly to the branches of some trees which stood near.

In the meantime they had left Robin by himself, thinking he would be safe while the gardener was mowing some grass; but what was the grief of both father and mother, when they returned, and could neither see nor hear him! The gardener, too, was gone; they therefore apprehended that a cat or rat had taken Robin away and killed him, yet none of his feathers were to be seen. With the most anxious search they explored every recess in which they thought it possible for him to be, and strained their little voices till they were hoarse with calling him, but all in vain. The tool-house was locked, but had he been there he would have answered: at length, quite in despair of finding him, with heavy hearts they returned to the nest; a general lamentation ensued, and this lately happy abode was now the region of sorrow. The father endeavoured to comfort his mate and surviving nestlings, and so far succeeded that they resolved to bear the loss with patience.

After a mournful night the mother left the nest early in the morning, unwilling to relinquish the hope which still remained of finding Robin again; but having spent an hour in this manner, she returned to her mate, who was comforting his little ones.

"Come," said he, "let us take a flight if we sit lamenting here for ever, it will be to no purpose: the evils which befall us must be borne, and the more quietly we submit to them, the lighter they will be. If poor Robin is dead, he will suffer no more: and if he is not, so much as we fly about, it is a chance but what we get tidings of him. Suppose these little ones attempt to fly with us to our benefactors? If we set out early, and let them rest frequently by the way, I think they may accomplish it." This was very pleasing to the little ones, and accordingly it was determined that they should immediately set out; they accomplished the journey by easy stages, and arrived in the court yard just after the daily pensioners were gone.

"Now," said the father, "stop a little, and let me advise you, Dicky, Flapsy, and Pecksy, to behave yourselves properly; hop only where you see your mother and me hop, and do not meddle with anything but what is scattered on purpose."

"Stay, father," said Dicky, "my feathers are sadly rumpled."

"And so are mine," said Flapsy.

"Well, smooth them then," said he; "but don't stand finicking for an hour."

Pecksy was ready in an instant, but the others were very tedious, so their father and mother would wait for them no longer, and flew in at the window; the others directly followed them, and, to the inexpressible satisfaction of Frederick, alighted on the tea-table, where they met with a very unexpected pleasure; for who should they find there as a guest but the poor lost Robin!

The meeting was, you may be sure, a happy one for all parties, and the transports it occasioned may be easier conceived than described. The father poured forth a loud song of gratitude, the mother chirped, she bowed her head, clapped her wings, basked on the tea-table, joined her beak to Robin's, then touched the hand of Frederick. The young ones twittered a thousand questions to Robin, but as he was unwilling to interrupt his father's song, he desired them to suspend their curiosity to another opportunity.

CHAPTER XI.

THE FEATHERED NEIGHBOURS.

You may remember, my young readers, that Frederick obtained from his mamma a promise that when the business of daily instruction was finished, he and his sister should go into the orchard in search of the robins. As soon, therefore, as the air was sufficiently cool she took them with her, and arrived just after the parent birds had taken their young ones back to the nest. Robin was then left by himself, and kept hopping about, and fearing no danger, got into the middle of the walk. Frederick descried him at a distance, and eagerly called out, "There's one of them, I declare!" and before his mamma observed him he ran to the place, and clapped his little hand over it, exulting that he had caught it. The pressure of his hand hurt Robin's wing, who sent forth piteous cries, on which Frederick let him go, saying, "I won't hurt you, you little thing."

Harriet, who saw him catch the bird, ran as fast as possible to prevent his detaining it, and perceived, as Robin hopped away, that he was lame, on which she concluded that her brother had hurt him; but on Frederick's assuring her that his wing hung down when he first saw him, Mrs. Benson said,—

"It is most likely he has been lamed by some accident, which has prevented his going with the others to the nest; and if that is the case, it will be humane and charitable to take care of him."

Frederick was delighted to hear her say so, and asked whether he might carry him home.

"Yes," said his mamma, "provided you can take him safely."

"Shall I carry him, madam?" said Joe; "he can lie nicely in my hat."

This was an excellent scheme, and all parties approved of it; so Frederick took some of the soft grass that was mowed down to put at the bottom, and poor Robin was safely deposited in this vehicle, which served him for a litter; and perceiving into what hands he had fallen, he inwardly rejoiced, knowing that he had an excellent chance of being provided for, as well as of seeing his dear relations again. I need not say that great care was taken of him, and you will easily suppose he had a more comfortable night than that he had passed in the shed.

When Frederick and Harriet arose the next morning, one of their first cares was to feed the birds, and they had the pleasure to see their nestlings in a very thriving condition; both the linnet and the blackbird now hopped out of their nests to be fed, to the great amusement of Frederick; but this pleasure was soon damped by an unlucky accident, for the blackbird being placed in a window which was open, hopped too near the edge, and fell to the ground, where he was snapped up by a dog, and torn to pieces in an instant. Frederick began to lament as before but on his sister's reminding him that the creature was past the sense of pain, he restrained himself, and turned his attention to the linnet, which he put into a cage, that he might not meet the same fate. He then went to feed the flock, and to inquire after Robin, whom Mrs. Benson had taken into her own room, lest Frederick should handle and hurt him. To his great joy he found him much better, for he could begin to use his injured wing; Frederick was therefore trusted to carry him into the breakfast parlour, where he placed him as has been already described.

For some time the young redbreasts behaved very well; but at length Dicky, familiarized by the kind treatment he met with, forgot his father's injunctions, and began to hop about in a very rude manner; he even jumped into the plate of bread and butter, and, wishing to taste the tea, hopped on the edge of a cup, but dipping his foot in the hot liquor, was glad to make a hasty retreat, to the great amusement of Frederick. Flapsy took the liberty of pecking at the sugar, but found it too hard for her tender beak. For these liberties. their mother reproved them, saying she would never bring them again, if they were guilty of such rudeness as to take what was not offered them.

As their longer stay would have broken in on a plan which Mrs. Benson had concerted, she rang her bell, and the footman came to remove the breakfast-things; on which the old birds, having taken leave of Robin, and promised to come again the next day, flew out at the window, followed by Dicky, Flapsy, and Pecksy. Robin was safely deposited in a cage, and passed a happy day, being often allowed to hop out in order to be fed.

The parent birds alighted in the court, and conducted their little ones to the water which was set out for them, after which they all returned to the nest; here the young ones rested till the afternoon, and then their parents took them out in order to show them the orchard.

"You have not yet seen," said the father, "the whole extent of this place, and I wish to introduce you to our neighbours."

He then led the way to a pear-tree, in which a linnet had built her nest. The old linnets seemed much pleased to see their friends the redbreasts, who with great pride introduced their little family to them.

"My own nestlings are just ready to fly," said the hen linnet, "and I hope will make acquaintance with them; for birds so well instructed, as I make no doubt your offspring are, must be very desirable companions."

The little redbreasts were delighted with the hopes of having some agreeable friends; and the old ones replied that they had themselves received so much pleasure from social friendship that they wished their young ones to cultivate it.

They then flew on to a cherry tree, in which were a pair of chaffinches in great agitation, endeavouring to part one of their own brood and a young sparrow which were engaged in a furious battle; but in vain: neither of the combatants would desist till the chaffinch dropped dead to the ground. His parents were greatly shocked at this accident; on which the cock redbreast attempted to comfort them with his strains; but finding them deaf to his music, he begged to know the cause of the quarrel.

"Oh," answered the hen chaffinch, "my nestling is lost through his own folly. I cautioned him repeatedly not to make acquaintance with sparrows, knowing they would lead him into mischief; but no remonstrances would prevail. As soon as he began to peck about, he formed a friendship with one of that voracious breed, who undertook to teach him to fly and provide for himself; so he left his parents, and continually followed the sparrow, who taught him to steal corn and other things, and to quarrel with every bird he met: I expected to see him killed continually. At length his companion grew tired of him, and picked a quarrel, which ended as you have seen. However, this is better than if he had been caught by men and hung up, as I have seen many a bird, for a spectacle, to deter others from stealing. Let me advise you, my young friends," said she, addressing herself to the young redbreasts, "to follow your parents' directions in every respect, and avoid bad company."

She then, accompanied by her mate, flew back to

her nest, in order to acquaint the rest of the family with this dreadful catastrophe, and the redbreasts took another flight.

They alighted on the ground, and began pecking about, when all of a sudden they heard a strange noise, which rather alarmed the young ones. Their father desired them to have no fears, but to follow him. He led them to the top of a high tree, in which was a nest of magpies, who had the day before made an excursion round the orchard, and were conversing on what they had seen, but in such a confused manner that there was no understanding them; one chattered of one thing, and one of another: in short, all were eager to speak, and none inclined to hear.

"What a set of foolish, ill-bred little creatures are these!" said the cock redbreast. "If they would talk one at a time, what each says might afford entertainment to the rest; but by chattering all together in this manner they are quite disagreeable. Take warning from them, my nestlings, and avoid the fault which renders them so ridiculous."

So saying, he flew on, and they soon saw a cuckoo surrounded by a number of birds, who had been pecking at her till she had scarce a feather left upon her breast, while she kept repeating her own dull note,

"Cuckoo! cuckoo!" incessantly. "Get back again to your own country," said a thrush; "what business have you in ours, sucking the eggs and taking the nests of any bird you meet with? Surely it might be sufficient that you have the privilege of building for yourself, as we do who are natives; but you have no right to seize upon our labours and devour our offspring."

"The cuckoo deserves her fate," said the hen redbreast: "though I am far from bearing enmity to foreign birds in general, I detest such characters as her. I wonder mankind do not drive cuckoos away; but I suppose it is on account of their being the harbingers of summer. How different is the character of the swallow! he comes here to enjoy the mildness of the climate, and confers a benefit on the land by destroying many noxious insects. I rejoice to see that race sporting in the air, and have had high pleasure in conversing with them; for, as they are great travellers, they have much to relate. But come, let us go on."

They soon came to a hollow tree. "Peep into this hole," said the cock bird to his young ones. They did so, and beheld a nest of young owls. "What a set of ugly creatures!" said Dicky; "surely you do not intend to show your frightful faces in the world!

Did ever any one see such dull eyes, and such a frightful muffle of feathers?"

"Whoever you are that reproach us with the want of beauty, you do not show your own good sense," replied one of the little owls; "perhaps we may have qualities which render us as amiable as yourselves. You do not appear to know that we are night, and not day birds; the quantity of feathers in which we are muffled up is very comfortable to us when we are out in the cold; and I can show you a pair of eyes which, if you are little birds, will frighten you out of your wits." He then drew back the film which was given him that the strong light of day might not injure his sight, and stared full at Dicky, who was struck with astonishment.

At that instant the parent owl returned, and seeing a parcel of strangers looking into her nest, she set up a screeching, which made the whole party take wing. As soon as they stopped to rest, the cock redbreast, who was really frightened, as well as his mate and family, recollected himself, and said, "Well, Dicky, how did you like the owl's eyes? I fancy they proved brighter than you expected; but had they even been as ugly as you supposed, it was very rude and silly in you to notice it. You ought never to censure any bird for natural deformities, since no one contracts them by choice; and what appears disagreeable to you may be pleasing in the eyes of another. Besides, you should be particularly careful not to insult strangers, because you cannot know their deserts, nor what power they may have of revenging themselves. You may think yourself happy if you never meet one of these owls by night, for I assure you they often feed upon little birds like us, and you have no reason to think they will spare you after the affront you have given them. But come, let us fly on." However, before we give any further account of their adventures, let us return to their benefactors.

CHAPTER XII.

THE VISIT TO THE FARM.

Just as Mrs. Benson and her children were preparing to leave the room, after having witnessed the happy meeting of the redbreast family at their tea-table, the servant entered and informed them that a poor woman was at the gate, who had been ordered to attend in the morning. Mrs. Benson desired she might come up. "Well, my good woman," said the benevolent lady, "how does your husband do?" "Thanks to your goodness, madam, and the blessing of God, quite cheery," said the woman.

"I am happy," said the lady, "to find you in better spirits than you were the other night, and do not doubt you will do very well. I will order some meat and bread to be sent you every day this week, and will also assist you in clothing the children."

Harriet's eyes glistened with benevolence at seeing the woman, whose distress had so greatly affected her, thus comforted; and slipping her purse, which contained seven shillings, into her mamma's hand, she begged her to take it for the woman. "You shall have the pleasure of relieving her yourself, my dear," said Mrs. Benson; "give this half-crown to her." Harriet, with a delight which none but the compassionate can know, extended the hand of charity. The woman received the benefaction with grateful acknowledgments, and praying that the Almighty might shower down his choicest blessings on this worthy family, respectfully took leave and returned to her husband, who, by means of the nourishment Mrs. Benson supplied her with, gathered strength hourly.

As soon as she was gone, Mrs. Benson informed her son and daughter that she intended to take them with her to Farmer Wilson's, where she made no doubt they would pass a happy day; and desired them to go and get ready for the journey while she dressed herself. The young folks obeyed without hesitation; and having given their maid very strict injunctions to feed Robin and the linnet, they attended their mamma to the carriage. Leaving this happy party to enjoy their pleasant drive, let us go back to the robins, whom we left on the wing in search of further adventures.

They soon alighted on a tree in which was a Mocking-bird,* who, instead of singing any note of his own, kept successively imitating those of every bird that inhabited the orchard, and this with a view of making them ridiculous. If any one had any natural imperfection in his singing, he was sure to mimic it, or if any one was particularly attentive to the duties of his station, he ridiculed him as grave and formal. The young redbreasts were excessively amused with this droll creature; but their father desired them to consider whether they should like to hear him mimic them. Every one agreed that they should be very angry to be ridiculed in that manner. "Then," replied the father, "neither encourage nor imitate him." The Mocking-bird, hearing him, took up his notes—"Neither encourage nor imitate him," said he. The cock redbreast on this flew at him with fury, plucked some feathers from his breast, and sent him screaming from the place. "I have made you sing a natural song at last," said he, "and hope you will take care how you practise mimicry again." His mate was sorry

* The Mocking-bird is properly a native of America, but is introduced here for the sake of the moral.

to see him disturb his temper and ruffle his feathers for such an insignificant creature; but he told her it was particularly necessary as an example to his nestlings, as mimicry was a fault to which young birds were too apt to incline, and he wished to show them the danger they exposed themselves to in the practice of it.

The whole redbreast family rested themselves for some time, and whilst they sat still they observed a chaffinch flying from tree to tree, chattering to every bird he had any knowledge of; and his discourse seemed to affect his hearers greatly, for they perceived some birds flying off in great haste, and others meeting them; many battles and disputes ensued. The little redbreasts wondered at these circumstances; at length Pecksy inquired the meaning of the bustle. "This chaffinch," replied the father, "is a tell-tale; it is inconceivable the mischief he makes. Not that he has so much malice in his nature, but he loves to hear himself chatter; and therefore every anecdote he can collect he tells to all he meets, by which means he often raises quarrels and animosities; neither does he stop here, for he frequently invents the tales he relates."

As the redbreast was speaking, the chaffinch alighted on the same tree. "Oh, my old friend," said he, "have you got abroad again? I heard the linnet in the pear tree say you were caught stealing corn, and hung up as a spectacle, but I thought this could not be true; besides, the blackbird in the cherry tree told me that the reason we did not see you as usual was that you were rearing a family, to whom, he said, you were so severe that the poor little creatures had no comfort of their lives."

"Whatever you may have heard, or whatever you may say, is a matter of indifference to me," replied the redbreast; but, as a neighbour, I cannot help advising you to restrain your tongue a little, and consider, before you communicate your intelligence, whether what you are going to say has a tendency to disturb the peace of society."

Whilst he was thus advising him, a flock of birds assembled about the tree; it consisted of those to whom the chaffinch had been chattering, who, having come to an explanation with each other, had detected his falsities, and determined to expel him from the orchard, which they did with every mark of contempt and ignominy. All the redbreasts joined in the pursuit, for even the little ones saw his character in a detestable light, and formed a determination to avoid his fault. When the tell-tale was gone, the party which pursued him alighted altogether in the same walk, and amongst them the redbreasts discovered many of their old friends, with whom they now renewed their acquaintance, knowing they should soon be released from family cares; and the young ones passed a happy day in this cheerful assembly. But at length the hour of repose approached, when each individual fled to his resting-place; and the redbreasts, after so fatiguing a day, soon fell asleep.

While the redbreasts were exploring the orchard, Mrs. Benson and her family, as we before showed, set off on their visit to the farm, where they met with a most welcome reception. Farmer Wilson was a very worthy, benevolent man. He had, by his industry, acquired sufficient to purchase the farm he lived on, and had a fair prospect of providing for a numerous family, whom he brought up with the greatest care, as farmers' sons and daughters used formerly to be, and taught them all to be merciful to the cattle which were employed in his business. His wife, a most amiable woman, had received a good education from her father, who was formerly schoolmaster of the parish. This good man had strongly implanted in his daughter's mind the Christian doctrine of universal charity, which she exercised, not only towards the human species, but also to poultry and every living creature which it was her province to manage. Mrs. Benson knew that her children would here have an opportunity of seeing many different animals treated with propriety; and it was on this account that she took them with her, though she herself visited these good people from a motive of sincere respect.

As soon as they were seated, Mrs. Wilson regaled her young guests with a piece of nice cake, made by her daughter Betsy, a little girl of twelve years old, who sat by, enjoying with secret delight the honour which the little lady and gentleman did to her performance. It happened fortunately to be a cool day, and Mrs. Benson expressed a desire to walk about and see the farm.

In the first place, Mrs. Wilson showed her the house, which was perfectly neat and in complete order. She then took her guests into her dairy, which was well stored with milk and cream, butter and cheese. From thence they went to visit the poultry-yard, where the little Bensons were excessively delighted, for there were a number of cocks and hens, and many broods of young chickens, besides turkeys and guinea-fowls.

All the fowls expressed the greatest joy at the sight of Mrs. Wilson and her daughter Betsy: the cocks celebrated their arrival by loud and cheerful crowings; the hens gave notice of their approach by cackling, and assembled their infant train to partake of their bounty; the turkeys and guinea-fowls ran to meet them; a number of pigeons also alighted from a pigeon-house. Betsy scattered among them the grain which she carried in her lap for that purpose, and seemed to have great pleasure in distributing it.

When their young visitors were satisfied with seeing the poultry fed, Mrs. Wilson showed them the henhouse and other conveniences provided for them, to make their lives comfortable; she then opened a little door which led to a meadow, where the fowls were often allowed to ramble and refresh themselves. On seeing her approach this place, the whole party collected, and ran into the meadow, like a troop of schoolboys into their playground.

"You, Mrs. Wilson, and your daughter, must have great amusement with these pretty creatures," said Mrs. Benson. "We have indeed, madam," said she, and they furnish us with eggs and chickens, not only for our own use, but for the market also."

"And can you prevail on yourself to kill these sweet creatures?" said Harriet. "Indeed, miss, I cannot," said Mrs. Wilson, "and never did kill a chicken in my life; but it is an easy matter to find people capable of doing it, and there is an absolute necessity for some of them to die, for they breed so fast that in a short time we should have more than we could possibly feed. But I make it a rule to render their lives as happy as possible; I never shut them up to fatten any longer than I can help, use no cruel methods of cramming them, nor do I confine them in a situation where they can see other fowls at liberty; neither do I take the chickens from the hen till she herself deserts them, nor set hens upon ducks' eggs."

"I often regret," said Mrs. Benson, "that so many lives should be sacrificed to preserve ours; but we must eat animals, or they would at length eat us— at least, all that would otherwise support us."

While this conversation passed, Frederick had followed the fowls into the meadow, where the turkey-cock, taking him for an enemy, had attacked him, and frightened him so much that he at first cried out for help, but soon recollected that this was cowardly; so he pulled off his hat, and drove the creature away before Betsy Wilson arrived, who was running to his assistance.

CHAPTER XIII.

THE PIGS AND BEES.

THE farmer's wife next proposed (but with many apologies for offering to take them to such a place) to show them her pigsty. The name of a pigsty generally conveys an idea of nastiness; but whoever had seen those of Farmer Wilson would have had a very different one. They were neatly paved, and washed down every day; the troughs in which the pigs fed were scoured, and the water they drank was always sweet and wholesome. The pigs themselves had an appearance of neatness which no one could have expected in such kind of animals; and though they had not the ingenuity of the Learned Pig, there was really something intelligent in their gruntings, and a very droll expression in the eyes of some of them. They knew their benefactors, and found means of testifying their joy at seeing them, which was increased when a boy, whom Mrs. Wilson had ordered to bring some bean-shells, emptied his basket before them. Now a scramble took place, and each pig began pushing the other aside and stuffing as fast as he could, lest they should have more than himself.

Harriet said she could not bear to see such greediness. "It is indeed very disagreeable, even in such creatures as these," replied Mrs. Benson, "but how much more so in the human species! and yet how frequent is this fault, among children in particular! Pray look at these pigs, Frederick, and tell me if you ever remember to have met with a little boy who ate strawberries as these pigs do bean-shells?" Frederick's cheeks, at this question, were covered with conscious blushes; on which his mamma kindly kissed him, and said she hoped he had seen enough of greediness to-day to serve him for a lesson as long as he lived.

In a separate sty was a sow with a litter of young pigs. This was a very pleasing sight indeed to Frederick, who longed to have one of them to play with; but Mrs. Wilson told him it would make the sow very angry, and her gruntings would terrify him more than the turkey-cock had done; on which he withdrew his request, but said he should like to keep such a little creature.

"If it would always continue little, Frederick," said Mrs. Benson, "it would do very well; but it will perhaps grow as large as its mother, and what should we do then?"

"I fear, ladies," said Mrs. Wilson, "you will be tired with staying here; will it be agreeable to you to take a walk in the garden?" "With all my heart," said Mrs. Benson.

Mrs. Wilson then conducted her guests into a garden, which abounded with all kinds of vegetables for the table, quantities of fruit, and a variety of flowers. Frederick longed to taste some of the delicacies which presented themselves to his eye, but he had been taught never to gather fruit or flowers without leave, nor ask for any. However, Mrs. Wilson, with his mamma's permission, treated him and his sister with some fine cherries, which Betsy gathered and presented in cabbage-leaves, and then took them to a shady arbour, where they sat and enjoyed their feast; after which they went to see the bees, which were at work in glass hives. This was a great entertainment, not only to the children, but to Mrs. Benson also, who was excessively pleased with the ingenuity and industry with which these insects collect their honey and wax, form their cells, and deposit their store. She had, by books, acquired a knowledge of the natural history of bees, which enabled her to examine their work with much greater satisfaction than she would have received from the sight of them had she been only taught to consider them as little stinging creatures which it was dangerous to approach.

"This is quite a treat to me indeed," said she to Mrs. Wilson, "for I never before had an opportunity of seeing bees work in glass hives."

"I find my account," said Mrs. Wilson, "in keeping bees thus, even upon a principle of economy; for as I do not destroy them, I have great numbers to work for me, and more honey every year than the last, notwithstanding I feed my bees in the winter. I have made acquaintance with the queen of every hive, who will come to me whenever I call her, and you shall see one of them if you please."

On this she called in a manner which the inhabitants of the hive they were looking at were accustomed to, and a large bee soon settled on her hand; in an instant after she was covered from head to foot with bees.

Harriet was fearful lest they should sting, and Frederick was running away; but Mrs. Wilson assured them the little creatures would not do any mischief if no one attempted to catch them. " Bees are, in their natural dispositions, very harmless creatures, I assure you, Master Benson," said she; "though I own they will certainly sting little boys who endeavour to catch them in order to suck their bags of honey or take out their sting; but you see that though I have hundreds about me, and even on my face and arms, not one offers to do me an injury; and I believe wasps seldom sting but in their own defence." She then threw up her hand, on which the queen bee flew away in great state, surrounded by her guards, and followed by the rest of her subjects, each ready to lose his own life in defence of hers.

"There is something very wonderful," said Mrs. Benson, "in the strong attachment these little creatures have to their sovereign, and very instructive too. What say you, Frederick, would you fight for your Queen?"

"Yes, mamma, if papa would," replied Frederick.

"That I assure you, my dear, he certainly would do if there were occasion, as loyally as the best bee in the world: and I beg you will remember what I now tell you as long as you live, that it is your duty to love your Queen, for she is to be considered as the mother of her people. But before we take our leave of the bees, let me observe to you, my dears, that several instructive lessons may be taken from their example. If such little insects as these perform their daily tasks with so much alacrity, surely it must be a shame for children to be idle, and to fret because they are put to learn things which will be of the utmost consequence to them in the end, and which would indeed conduce to their present happiness, would they but apply to them with a willing mind. Remember the pretty hymn you have learnt,—

> 'How doth the little busy bee
> Improve each shining hour!' &c.

But come, Mrs. Wilson," continued the lady, "we must, if you please, take leave of the bees, or we shall have no time to enjoy the other pleasures you have in reserve for us."

As they walked along, Frederick so far forgot himself as to try to catch a moth, but his mamma obliged him to let it go immediately. "Don't you think, Mrs. Wilson," said she, "it is wrong to let children catch butterflies and moths?" "Indeed I do, madam," replied the good woman. "Poor little creatures! what injury can they do us by flying about? In that state, at least, they are harmless to us. Caterpillars and snails, it is true, we are obliged frequently to destroy, on account of their devouring fruit and vegetables; but unless they abound so as to be likely to do a real injury, I never suffer them to be meddled with. I often think on my good father's maxims, which were—'Never take away the life of any creature, unless it is necessary for the benefit of mankind. While there is food and room enough in the world for them and us, let them live and enjoy the blessings they were formed for,' he would say."

"When I was a little girl," said Mrs. Benson, "I had a great propensity to catch flies and other insects; but my father had an excellent microscope, in which he showed me a number of different objects; by this means I learnt that even the minutest creatures might be as susceptible of pain as myself. And so far from having a pleasure in killing even the disagreeable insects which are troublesome in houses, I assure you I cannot do it or see it done without pain; and yet they certainly may be considered as enemies, and as such we have a right to destroy them."

"To be sure, madam," said Mrs. Wilson, "for without cleanliness we could not enjoy health. It goes against me to demolish a fine spider's web, and yet they make a house look very dirty; but I seldom have any in mine, for I took care, when I first came to live in it, to destroy the nests, and the old spiders finding there was no security for their young ones here, have forsaken the house; and I am inclined to think the same vigilance in respect to other disagreeable insects would have the same effect."

"Doubtless," said Mrs. Benson; "but pray tell me, do you destroy the webs of garden spiders also?" "Not unless there are so many as to be disagreeable," replied Mrs. Wilson. "I should not myself like to have the fruits of my industry demolished, nor my little ones taken out of my arms, or from their warm beds, and crushed to death."

"I am of opinion," said Mrs. Benson, "that it would be a good way to accustom one's self, before one kills anything, to change situations with it in imagination, and to suppose how we should feel were we bees, or ants, or birds, or kittens, and so on."

"Indeed, madam," said Mrs. Wilson, 'I have often wished that poor dumb creatures had somebody to speak for them; many an innocent life would then be saved, which is now destroyed to no end."

"Well," said Harriet, "I am sure I shall never kill anything without first magnifying it in my mind, and thinking what it would say for itself if able to speak." "Then, my dear, I will engage for you," replied her mamma, "that you will put but very few creatures to death. But in order to have a proper notion of their form, you must study natural history, from whence you will learn how wonderful their construction is, how carefully and tenderly the inferior creatures provide for their young, how ingenious their various employments are, how far they are from harbouring malice against the human species, and how excellently they are informed and instructed by their great Creator for the enjoyment of happiness in their different classes of existence, which happiness we have certainly no right wantonly to disturb.

"Besides, it is really a meanness to destroy any creature merely because it is little, and, in children, particularly absurd to do so; for, upon this principle, they must themselves expect to be constantly ill-treated, though no animal stands more in need of tenderness than they do for many years from the time of their coming into the world: and even men and women might expect to be annihilated by the power of the great Creator, if everything that is little were to be destroyed. Neither do I know how we can precisely call anything great or little, since it is only so by comparing it with others. An ant or a fly may appear, to one of its own species, whose eyes are formed to see those parts which we cannot discover without glasses, as considerable as men and women do to each other; and to creatures of the dimensions of a mite, one of the size of an ant doubtless looks formidable and gigantic. I therefore think it but justice to view insects with microscopic eyes before we do anything to them that is likely to give them pain, or to destroy their works unnecessarily."

During this conversation Frederick kept running about, making choice of flowers, which Betsy Wilson formed into nosegays for his mamma, his sister, and himself.

FREDERICK VIEWING THE DUCKS AND GEESE.

CHAPTER XIV.

FREDERICK VIEWING THE DUCKS AND GEESE.

THE next place Mrs. Wilson took her guests to was a barnyard, in which was a large horse-pond. Here her young visitors were delighted with the appearance of a number of geese and ducks; some were swimming in the water, some diving, others routing in the mud to see what fish or worms they could find.

"It appears very strange to me," said Harriet, "that any creatures can take delight in making themselves so dirty." "And yet," replied Mrs. Benson, "how many children do the same, without having any excuse for it! The ducks and geese grub about so in search of the necessaries of life; but I have seen boys do it merely for diversion, and sometimes at the hazard of their lives."

"Have you any fish here?" said Frederick. "I believe none of any consequence," replied Mrs. Wilson; "the ducks and the geese would take care that none should grow to any considerable size. But there are plenty in a pond which you will see in the next field, and I hope to have the pleasure of seeing you, at dinner, eat of some perch which were caught there. Sometimes we catch fine carp and tench, but only with nets; for neither my good man nor I can bear the cruel diversion of angling, nor do we allow our children to follow it, from a notion that it hardens the heart and leads to idleness.

"Pray, mamma," said Harriet, "is it right to catch fish? I should think, as they live in water, and we upon land, we have no business with them." "You would wish every one, then, my dear, to keep to their own element? Your sentiment is a good one in many respects, but it must not be extended so far as to forbid the catching of fish. Man has dominion over the fish, as well as over beasts and fowls, and many of them are proper food for mankind, and the astonishing increase of them shows that they are designed to be so; for were all that are spawned to grow to full size, there would soon be more than our ponds, or even than the sea itself, could hold, and they would be starved: therefore there are the same reasons for our feeding upon them as upon poultry, but we should be very careful to despatch them as quickly as possible.

"Some people are cruel enough to roast lobsters alive, the cries of which, I have been told, are dreadful to hear; and others will flay eels alive, then put them without their skins into a pail of cold water, and afterwards cut them in pieces, and throw them into a frying-pan of boiling fat, where sometimes every separate piece will writhe about in agony: thus each poor fish suffers as many deaths as it is divided into pieces. Now this, Harriet, cannot be right, however authorized by custom; therefore I hope you will never suffer such things to be done in your kitchen when you keep house, but always give orders that your lobsters be put into boiling water, which kills them soon, and that your eels be killed before they are skinned, which may soon be done by laying hold of their heads and tails and giving them a sudden pull, which separates the vertebræ of the back. This is dreadful enough, though little in comparison of what they suffer by the other methods of killing them."

"Oh, mamma!" said Harriet, "you make me even shudder! I do not believe I shall ever desire to eat eels; I shall be ready to make speeches for every piece as it lies in a dish before me. But pray tell me, is it cruel to kill frogs and toads?" "Ask Mrs. Wilson, my dear; she has more to do with such reptiles than I have," said Mrs. Benson. "Well," replied Mrs. Wilson, "I am very singular in regard to such kind of creatures, and though I by no means like to have them in my house, I do not make an outcry, and condemn every one to a violent death which is accidentally found in my cellars or other places: on the contrary, I generally see them thrown into a ditch at some distance, to take their chance. There are many birds and water-fowl that feed on young frogs and toads, which will in general keep them from multiplying so as to be nuisances to us, and it is time enough for us to take arms against them if there happen to be a very extraordinary increase of them. My good man is as particular in respect to moles: if he finds them in his garden, or any other part of his grounds where they can do mischief, he has them killed, but never suffers them to be molested when they are harmless. Neither does he hunt after snakes, or permit any one belonging to him to do so; for he says that if they are not disturbed they will not come from their haunts to annoy us; and to kill for the sake of killing is cruel."

"Pray, Mrs. Wilson," said Frederick, "do your sons ever go bird's-nesting?" "No," said she; "I hope I have not a child amongst my family capable of such barbarity. In the course of the summer they generally have young birds to nurse, which fall out of their nests or lose their parents, but are seldom lucky enough to raise them, and we have only one in a cage which they reared last summer. Yet we have plenty of singing, for the sweet creatures, finding they may enjoy themselves unmolested in the trees, treat us with their harmony from morning to night, of which you had a specimen in the garden. Sparrows, indeed, my husband is under the necessity of destroying, for they are such devourers, they would leave him but little corn to carry to market if he did not shoot them: but he never kills the crows, because they are very serviceable in picking up grubs and other things injurious to farmers. We only set a little boy to watch our new-sown grain, and he keeps making a noise, which effectually frightens them."

"Oh," said Frederick, "I nurse young birds too. I have got a linnet and a robin redbreast, and I feed a hundred beside." Mrs. Wilson smiled, and addressing herself to Mrs. Benson, said, "Now, madam, we will, if you please, return to the house, for I fancy by this time dinner is nearly ready, and my husband and sons are about coming home."

Mrs. Benson was a little tired with her ramble, and was really impatient to see Farmer Wilson and the rest of his amiable family. When she drew near the house, she was met by the worthy man, who gave her a most cordial welcome, and said he was proud to see so much good company. Nancy, the eldest daughter, to whom the mother had entrusted the care of inspecting the additional cookery which she had ordered, and who, for that reason, was not to be seen in the morning, now made her appearance, dressed with the most perfect neatness; health bloomed in her cheeks, and cheerfulness and good humour sparkled in her eyes. With this engaging countenance she easily prevailed on Frederick to let her place him by her at the table, round which the two other visitors, the master and mistress of the house, and the rest of their offspring, consisting of Thomas, a fine youth of eighteen, four young boys, and little Betsy, were soon seated.

The table was covered with plain food, but, by the good management of Nancy, who had made an excellent pudding, an apple pie, and some delicious custards, it made a very good figure; and Mrs. Benson afterwards declared that she had never enjoyed an entertainment so much : and the pleasure was considerably brightened by the happy countenances of the whole family.

The farmer, who was a jocose man, said a number of droll things, which diverted his little visitors very much ; and soon after dinner he begged leave to depart, as he was sheep-shearing, but said he thought the young gentlefolks might be diverted with the sight, so invited them to pay a visit in the field, and left Joe and Neddy to conduct Frederick to it. The young farmers were rather shy at first, being afraid that their guests would laugh at their country talk ; but when they observed how politely they behaved to their sisters, they entered into conversation, and told Frederick a hundred particulars about animals, with which he was before unacquainted; and he in return related all he knew about his redbreasts and other pensioners. They then showed him a pretty cat with kittens, and also their favourite, Daphne, with two young puppies : the latter were kept in a kennel, and the cat in a stable, where they were well supplied with food.

As Frederick knew that his sister was remarkably fond of cats, he stepped back to call her to look at them, which, with her mamma's permission, she was greatly pleased to do, and longed to have the kittens to nurse. When she returned, she inquired whether the dogs and cats were ever permitted to come into the house.

"Not whilst they have young ones," said Mrs. Wilson, "for they make a great deal of dirt, and are very troublesome at that time. But when Puss has brought up her family, which is designed for the stable, she shall be admitted amongst us again, for she is a very useful creature, and deserves to be well treated. But I do not suffer my children to handle her; I think it looks very ugly for any one to be all over scratches. Daphne is admitted to a greater share of familiarity; she is very faithful and extremely good-natured; but we never feed her in the house, for there is no doing so without greasing the floors."

"I am opinion," said Mrs. Benson, "that a difference should be made between our treatment of cats and of dogs. There is something very savage in the nature of the former; and though they certainly are deserving of our kindness on account of their usefulness, yet they cannot make themselves so agreeable as dogs; and there is really something very formidable in their talons and teeth, and when enraged a cat is no better than a little tigress. Besides, were there not danger to one's self in nursing cats, there is no doing it without injury to one's linen, for when Puss is best pleased she generally tramples with her talons unsheathed, by which practice many a dress has been torn. And even the cleanliness of cats is injurious, for they usually have recourse to the corners of chairs in order to rub the dirt from their talons. Many people have a great dread of this animal, and on that account it should not be used to come into rooms in which a variety of company is received. As for dogs, they are in general so very social, grateful, and pleasing, that they seem formed to be the humble companions of mankind, and if kept in proper order, they may be familiarized with safety; but then they should be taught to know their distance. And as there are different species of dogs, we should make a prudent selection, and not introduce into the house great mastiffs or tall greyhounds; neither must we indulge those we domesticate to too great a degree, for in that case they will become as troublesome as cats."

Mrs. Benson now expressed her desire to see the sheep-shearing, on which Mrs. Wilson and her daughter conducted her and Harriet to the field, where they arrived at the conclusion of the operation And a very pleasing sight it was to behold the happy creatures, who lately waddled under a heavy,

M

heating load, relieved from their burden, leaping and frisking with delight, whilst the accumulated wool seemed, as it lay, to promise comfortable clothing for many a naked person among the human species, who, destitute of such a supply, would be in danger of perishing in the ensuing winter.

Harriet observed the innocent countenances of the sheep and lambs, and said she thought it was a thousand pities to kill them. "It is so, my dear," said her mamma; "but we must not indulge our feelings too far in respect to animals which are given us for food; all we have to do is to avoid barbarity. It is happy for them that, having no apprehension of being killed, they enjoy life in peace and security to the very last, and even when the knife is lifted to their throats, they are ignorant of its destination, and a few struggles put an end to their pain for ever. But come, Mrs. Wilson, will you favour us with a sight of your cows?" "With pleasure, madam," said she; "they are by this time driven up to be milked."

Mrs. Wilson then conducted her visitors towards the farmyard. "Perhaps, madam," said she, as they walked along, "the young lady and gentleman may be afraid of horned cattle?" "I believe," replied Mrs. Benson, "I may venture to say that Harriet has no unreasonable fears of any living creature; it has been my endeavour to guard the minds of my children against so distressing a weakness: but whether Frederick's heart has acquired fortitude enough to enable him to venture near so many cows I cannot tell. "Oh yes, mamma," cried Frederick. "I would sooner get up and ride into the yard on the horns of one of them than run away." "Well, we shall soon put your courage to the proof," said Mrs. Benson; "so come along, sir."

"As for my children," said Mrs. Wilson, "they are remarkably courageous in respect to animals. All the creatures belonging to us are very harmless and gentle, which is the natural consequence of kind treatment, and no person need be afraid of walking in any part of our grounds; but it is difficult to persuade some people that there is no danger, for they are apt to imagine that every loose horse they see will gallop over them, and that every creature with horns will gore and toss them."

"Very true," replied Mrs. Benson; "and I have known many as much afraid of a toad, a frog, or a spider, as if certain death would be the consequence of meeting them; when, if these persons would but make use of their reason, they would soon be convinced that such fears are ill-grounded. Frogs and toads are very harmless creatures, and so far from offering an injury to any human being they may chance to meet, they hop away with all possible expedition, from a dread of being themselves destroyed; and spiders drop suddenly down with a view to their own preservation only; and therefore it is highly ridiculous to be afraid of them.

"Horses and oxen are much more formidable creatures; they certainly could do us a great deal of mischief if they were conscious of their superior strength; but God has wisely ordained that they should not be so, and having given mankind dominion over them, He has implanted in their nature an awe and dread of the human species, which occasion them to yield subjection to the lords of the creation, when they exert their authority in a proper manner. It is really a very wonderful thing, Mrs. Wilson, to see a fine lively horse submitting to the bit and harness, or a drove of oxen quietly marching under the direction of one man. But it is observable that those creatures which are the most useful to us are the easiest tamed, and yield, not only singly, but in flocks, to mankind—nay, even to boys. This shows at once the goodness and power of the Creator. From what I have said, my dears," added Mrs. Benson, " you must perceive that it is a great weakness for a human being to be afraid of animals."

By this time the party were advanced pretty near to the farmyard, and Frederick espied one of the cows peeping over the gate; on which, with a countenance expressive of fear, he ran hastily to his mamma, and asked her whether cows could toss people over gates and hedges. "What a silly question, Frederick!" said she: "pray look again, and you will perceive that it is impossible for such large heavy creatures to do so; and these enclosures are made on purpose to confine them within proper bounds. But did you not boast just now that 'you could ride on the horns of one of them'? That I shall not require you to do, for it would very likely make the creature angry, because cows are not accustomed to carry any load upon their heads; neither would I allow you to run after them with a stick, or to make any attempt to frighten them; but if you approach as a friend, I make no doubt you will be received as such. So summon your courage, and attend us; the cows will not hurt you, I can assure you."

Neddy Wilson then began laughing, from the idea that a boy should be afraid of a cow, which made Frederick ashamed of himself and quitting his mamma's gown, by which he had held fast while she was speaking, he laid fast hold of Neddy's hand, and declared his resolution to go as near the cow as he would. I will not take upon me to say that his little heart was perfectly free from palpitation, but that lay in his own bosom, where none could discover its feelings but himself; so let us give him as much credit for courage as we can, and acknowledge him to have been a noble little fellow in thus trusting himself amongst a number of horned cattle.

The whole party now entered the farmyard, where they saw eight fine cows, fat, sleek, and beautifully clean, who yielded several pails of rich milk, the steam of which, added to the breath of the cows, cast a delightful fragrance around. Mrs. Wilson then entreated her company to return to the house, where tea was provided, and a delicious syllabub.

The farmer now came back, and refreshed himself with a cup of ale, which was very comfortable after the fatigues of the day.

"I have had," said Mrs. Benson, "great pleasure in viewing your farm, Mr. Wilson, which appears to me to afford all the desirable comforts and conveniences of life, and I most sincerely wish a continuance of your prosperity. If it is not an impertinent question, pray tell me, did you inherit the farm from your father, or was it purchased with the fruits of your own industry?" "Neither my wife nor I have led an idle life, I assure you, madam," replied the farmer; "but, next to the blessing of Heaven, I think myself in a great degree indebted to my cattle for my good success. My father left me master of a little farm, with a few acres of land well cropped, three horses, two cows, ten sheep, a sow and pigs, an ass, and a few poultry: these have gradually multiplied to what you now see me possess, besides numbers that I have sold; and I have had fine crops of hay and corn, so that every year I laid by a little money, till I was able to purchase this farm, which has proved a very good one to me."

"There is something so uncommon in hearing a farmer attribute a part of his success in life to his cattle, that I should be obliged to you, Mr. Wilson," said the lady, "if you would account to me for this circumstance."

"Most readily, madam," said he. "When I was a very young man, I heard a fine sermon from the pulpit on the subject of showing mercy to brutes, which made a great impression on my mind; and I have ever since acted towards all dumb creatures as I would to mankind, upon the principle of doing as I would be done by. I always consider every beast that works for me as my servant, and entitled to wages; but as beasts cannot use money, I pay them in things of more value to them, and make it a rule, unless in case of great necessity, to let them enjoy rest on the Sabbath day.

"I am very cautious of not letting my beasts work beyond their strength, and always give them their food in due season; nor do I ever suffer them to be beat or cruelly used. Besides giving them what I call their daily wages, I indulge them with all the comforts I can afford them. In summer, when the business of the day is over, my horses enjoy themselves in a good pasture, and in winter they are sheltered from the inclemencies of the weather in a warm stable. If they get old, I contrive some easy task for them; and when they can work no longer, I let them live on the common without it, till age and infirmities make their lives burthensome to themselves, when I have them put to as easy a death as possible.

"Though my cows and sheep do not work for me, I think them entitled to a recompence for the profit I receive from their milk and wool, and endeavour to repay them with the kindest usage; and even my ass finds mercy from me, for I could not bear to see so useful a creature ill-treated; and as for my dogs, I set great store by them on account of their fidelity."

"These are very excellent rules indeed, Mr. Wilson, and I wish they were generally followed," said Mrs. Benson; " for I believe many poor beasts suffer greatly from the ill-treatment inflicted on them, the horses particularly." Yes, madam," said the farmer, "I have heard so, and could tell you such stories of cruelties exercised on brutes in the country as would quite shock you ; and I have seen such instances myself of the ill effects of neglecting them, as have confirmed me in the notions I learnt from the good sermon I told you of."

"I am much obliged to you for your information, Mr. Wilson," said Mrs. Benson, " and hope my children will never forget it, for it certainly is a duty to extend our clemency to beasts and other animals. Nay, we are strictly commanded in the Scriptures to show compassion to the beasts of others, even to those of our enemies ; surely, then, those which are our own property, and work for us, have a peculiar claim to it. There is one custom which shocks me very much, and that is pounding of cattle; I fancy, Mr. Wilson, you do not practise that much."
" Madam," replied he, "I should much rather pound the owners of them, through whose neglect or dishonesty it generally happens that horses trespass on other people's land. If any beast accidentally gets

into my grounds, I send it home to its owner, for it certainly is no wilful fault in the creature to seek the best pasture it can find; but if I have reason to suppose his owner turned him in, I then think myself bound to do what the law directs in that respect: but though it is a secret I am obliged to keep from my neighbours, I may safely confess to you, madam, that I have not the heart to let a poor beast starve in a pound. As there are no courts of justice in which beasts can seek redress, I set up one for them in my own breast, where humanity pleads their cause."

"I wish they had such an advocate in every breast, Mr. Wilson," said the lady; "but my watch reminds me we must now take our leave, which I do with many thanks to you and Mrs. Wilson for your kind entertainment and good cheer, and shall be happy to return your civilities at my own house, and pray bring your whole family with you."

Mrs. Benson then desired her son and daughter to prepare for their departure. Fredrick was grown so intimate with little Neddy that he could scarcely be prevailed on to leave him, till he recollected Robin and the linnet.

As they returned in the carriage, Mrs. Benson remarked that Farmer Wilson's story was enough to make every one who heard it careful of their live stock, for their own sakes; "But, said she, "the pleasure and advantage will be greatly increased if it is done from a principle of humanity as well as interest." Harriet answered that she hoped she should neither treat animals ill nor place her affections on them too strongly. "That, my dear," replied her good mamma, is the proper medium to be observed."

In a short time they arrived at home. The maid to whose care the birds had been entrusted gave a good account of her charge; and Harriet and Frederick went to bed in peace, after a day spent with much pleasure and improvement.

CHAPTER XV.

THE AVIARY.

THE next morning the redbreasts attended at Mrs. Benson's as usual, and Robin was still better, but his father began to fear he would never perfectly recover from his accident; however, he kept his apprehensions to himself, and suffered the little ones to entertain their lame brother with a relation of what they had seen the day before in the orchard. Frederick and Harriet were so diverted with the chattering and chirping of the little things, that they did not miss the parent's song.

When the young ones had stayed as long as she thought right, the hen redbreast summoned them away, and all took leave of Robin, who longed to go with them, but was not able. The father reminded him that he had great reason to rejoice in his present situation, considering all things; on which he resumed his cheerfulness, and giving a sprightly twitter, hopped into Frederick's hand, which was spread open to receive him. The rest then flew away, and Harriet and her brother prepared for their morning tasks.

The redbreasts alighted as usual to drink in the courtyard, and were preparing to return to the orchard, when Flapsy expressed a desire to look a little about the world, for she said it would be very mopish to be always confined to the orchard; and Dicky seconded her request. Pecksy declared that, however her curiosity might be excited, she had known so much happiness in the nest, that she was strongly attached to the paternal spot, and could gladly pass her life there. The parents highly commended her contented disposition; but her father said that as there was nothing blameable in the inclination Dicky and Flapsy expressed for seeing the world, provided it was kept within due bounds, he would readily gratify it. Then asking if they were sufficiently refreshed, he took wing, and led the way to a neighbouring grove, where he placed his little tribe among the branches of a venerable oak. Here their ears were charmed with a most enchanting concert of music. On one tree a blackbird and a thrush poured forth their strong melodious notes; on another a number of linnets joined their sweet voices; exalted in the air a skylark modulated his delightful pipe, whilst a brother of the wood, seated on a cool refreshing turf, made the grove re-echo with his melody; to these the nightingale joined his enchanting lay: in short, not a note was wanting to complete the harmony.

The little redbreasts were so exceedingly charmed, that for a while they continued listening with silent rapture. At length Dicky exclaimed, "How happy should I be to join the cheerful band, and live for ever in this charming place!" "It is," replied his mother, "a very pleasant situation, to be sure; but could you be sensible of the superior advantages which, as a redbreast, you may enjoy by taking up your abode in the orchard, you would never wish to change it. For my own part, I find myself so happy in that calm retreat, that nothing but necessity shall ever drive me from it."

Pecksy declared that though she was much delighted with the novelty of the scene, and charmed with the music, she now felt an ardent desire to return home; but Flapsy wished to see a little more first. "Well," said the father, "your desire shall be gratified; let us take a circuit in this grove, for I wish you to see everything worth observation in every place you go to, and not fly about the world, as many giddy birds do, without the least improvement from their travels." On this he spread his wings as a signal of departure, which his family obeyed.

Observing a parcel of boys creeping silently along, "Stop," said he, "let us perch on this tree, and see what these little monsters are about." Scarcely were they seated, when one of the boys mounted an adjacent tree, and took a nest of half-fledged linnets, which he brought in triumph to his companions.

At this instant a family of thrushes unfortunately chirped, which directed another boy to the place of their habitation, to which he climbed, and eagerly seized the unfortunate little creatures. Having met with so much success, the boys left the grove, to exult at their own homes over their wretched captives, for ever separated from their tender parents, who soon came back laden with the gain of their labour, which they had kindly destined for the sustenance of their infant broods.

The little redbreasts were now spectators of those parental agonies which had been formerly described to them, and Pecksy cried out, " Who would desire to live in this grove, after having experienced the comforts of the orchard?" Dicky and Flapsy were desirous to depart, being alarmed for their own safety. "No," said the father, "let us stay a little longer—now we will go on."

They accordingly took another flight, and saw a man scattering seed upon the ground. "See there," said Dicky, "what fine food that man throws down! I dare say he is some good creature who is a friend to the feathered race. Shall we alight and partake of his bounty?" "Do not form too hasty an opinion, Dicky," said the father; "watch here with me a little while, and then do as you will." All the little ones stretched their necks, and kept a curious eye fixed on the man. In a few minutes a number of sparrows, chaffinches, and linnets descended, and began to regale themselves; but, in the midst of their feast, a net was suddenly cast over them, and they were all taken captive. The man, who was a birdcatcher by profession, called to his assistant, who brought a cage divided into a number of small partitions, in which the linnets and chaffinches were separately deposited. In this dismal prison, where they had scarcely room to flutter, were those little creatures confined who lately poured forth their songs of joy fearless of danger. As for the sparrows, their necks were wrung, and they were put in a bag together. The little redbreasts trembled for themselves, and were in great haste to take wing. "Stay," said the father, "Dicky has not yet made acquaintance with this friend of the feathered race." "No," said Dicky, "nor do I desire it; defend me and all who are dear to me from such friends as these!" "Well," said the father, "learn from this instance never to form a hasty judgment, nor to put yourself in the power of strangers, who offer you favours you have no right to expect from their hands."

"Indeed, my love," said the mother bird, "I am very anxious to get home; I have not lately been used to be long absent from it, and every excursion I make endears it more to me." "Oh, the day is not half spent," replied her mate, "and I hope that for the gratification of the little ones you will consent to complete the ramble. Come, let us visit another part of the grove; I am acquainted with its inmost recesses." His mate acquiesced, and they proceeded on their journey.

At length the father hastily called out, "Turn this way! turn this way!" The whole party obeyed the word of command, and found the good effects of their obedience, for in an instant they saw a flash of fire, a thick smoke followed it, and immediately they heard a dreadful sound, and saw a young redstart fall bleeding to the ground, on which he struggled just long enough to cry, "Oh, my dear father! why did I not listen to your kind admonitions, which I now find, too late, were the dictates of tenderness!" and then expired.

The little redbreasts were struck with consternation at this dreadful accident, and Pecksy, who recovered the soonest, begged her father would inform her by what means the redstart was killed. "He was shot to death," said he, "and had you not followed my directions, it might have been the fate of every one of you; therefore let it be a lesson to you to follow every injunction of your parents with the same readiness for the future. You may depend upon it our experience teaches us to foresee many dangers which such young creatures as you have no notion of, and when we desire you to do or to forbear anything, it is for the sake of your safety or advantage. Therefore, Dicky, never more stand, as you sometimes have done, asking why we tell you to do so and so; for had that been the case now, you, who were in a direct line with the gunner, would have been inevitably shot."

They all said they would pay implicit obedience. "Do so," said he; "but in order to this you must also remember to practise in our absence what we enjoin you when present. For instance, some kinds of food are very prejudicial to your health, which we would not, on any account, let you taste when we are by; these you must not indulge in when away from us, whatever any other bird may say in recommendation of them. Neither must you engage in any dangerous enterprise, which others, who have natural strength or acquired agility, go through with safety; nor should you go to any places which we have pointed out as dangerous, nor join any companions which we have forbidden you to make acquaintance with. This poor redstart might have avoided his fate, for I heard his father, when I was last in the grove, advise him not to fly about by himself till he had shown him the dangers of the world."

Pecksy answered that she knew the value of parental instruction so well, that she should certainly treasure up in her heart every maxim of it; and the others promised to do the same. "But," said Flapsy, "I cannot understand the nature of the accident which occasioned the death of the redstart."

"Neither can I explain it to you, my dear," replied the father; "I only know that it is a very common practice with some men to carry instruments from which they discharge what proves fatal to many a bird; but I have, by attentive observation, learnt how to evade the mischief. But come, let us descend and rest ourselves a little, as we may do it with safety, and then we will see if we cannot find a place where you can find amusement, without being exposed to such dangers as attend the inhabitants of woods and groves. Are you sufficiently rested to take a pretty long flight?" "Oh yes," cried Dicky, who was quite eager to leave the spot in which, a short time before, he had longed to pass his life: the rest joined in the same wish, and every wing was instantly expanded.

The father led the way, and in a very short time he and his family arrived at the estate of a gentleman who, having a plentiful fortune, endeavoured to collect all that was curious in art and nature, for the amusement of his own mind and the gratification of others. He had a house like a palace, furnished with every expensive rarity; his gardens, to which the redbreasts took their flight, were laid out in such a manner as to afford the most delightful variety to the eye.

Amongst other articles of taste was an aviary, which was built like a temple, enclosed with brass wire. The framework was painted green, and ornamented with carving, gilt; in the middle a fountain continually threw up fresh water, which fell into a basin whose brink was enamelled with flowers; at one end were partitions for birds' nests, and troughs containing various kinds of seed and materials for building nests: this part was carefully sheltered from every inclemency of the weather. Numbers of perches were placed in different parts of the aviary, and it was surrounded by a most beautiful shrubbery.

A habitation like this, in which all the conveniences of life seemed to be collected, where abundance was supplied without toil, where each gay songster might sing himself to repose in the midst of ease and plenty, safe from the dangers of the woods, appeared to our young travellers desirable beyond all situations in the world, and Dicky expressed an earnest wish to be admitted into it. "Well," said the father, "let us not determine hastily; it will be advisable first to inquire whether its inhabitants are really happy, before you make interest to become one of the number. Place yourselves by me on this shrub, and whilst we rest ourselves we shall have an opportunity of seeing what passes."

The first bird that attracted their notice was a dove, who sat cooing by himself in a corner, in accents so gentle and sweet, that a stranger to his language would have listened to them with delight; but the redbreasts, who understood their import, heard them with sympathetic concern. "Oh, my dear, my beloved mate!" said he, "am I then divided from you for ever? What avails it that I am furnished here with all the elegances and luxuries of life? Deprived of your company, I have no enjoyment of them; the humblest morsel, though gained with toil and danger, would be infinitely preferable to me if shared with you. Here am I shut up for the remainder of my days, in society for which I have no relish, whilst she who has hitherto been the beloved partner of all my joys is for ever separated from me! In vain will you, with painful wing, pursue your anxious search in quest of me; never, never more shall I bring you the welcome refreshment; never shall I hear your soothing voice, and delight in the soft murmurs of the infant pair which you hatched with such care and nursed with such tenderness! No, my beloved nestlings, never will your wretched father be at liberty to guide your flight and instruct you in your duty." Here his voice faltered, and overcome with bitter reflections, he resigned himself a prey to silent sorrow.

"This dove is not happy, however," said the hen redbreast to her mate, "and no wonder; but let us attend to the notes of that lark." His eyes were turned up towards the sky, he fluttered his wings, he strained his throat, and would, to a human eye, have appeared in raptures of joy; but the redbreasts perceived that he was inflamed with rage. "And am I to be constantly confined in this horrid place?" sang he. "Is my upward flight to be impeded by bars and wires? Must I no longer soar towards that bright luminary, and make the arch of heaven resound with my singing? Shall I cease to be the herald of the morn, or must I be so in this contracted sphere? No, ye partners of my captivity, henceforth sleep on and take ignoble rest, and may you lose in slumber the remembrance of past pleasures! Oh, cruel and unjust man! was it not enough that I proclaimed the approach of day, that I soothed your sultry hours, that I heightened the delights of evening; but must I, to gratify your unfeeling wantonness, be secluded from every joy my heart holds dear, and condemned to a situation I detest? Take your delicious dainties, reserve your flowing stream, for those who can relish them, but give me liberty! But why do I address myself to you, who are heedless of my misery?" Here, casting an indignant look around, he stopped his song.

"What think you now, Dicky?" said the redbreast; "have you as high an idea of the happiness of this place as you conceived at the first view of it?"

"I cannot help thinking still," replied Dicky, "that it is a charming retreat, and that it must be very comfortable to have everything provided for one's use."

"Well," said the father, "let us move, and observe those linnets who are building their nest."

Accordingly, they flew to a tree, the branches of which formed a part of the shelter of the aviary, where they easily heard, without being themselves observed, all that passed in it.

"Come," said one of the linnets, "let us go on with our work and finish the nest, though it will rather be a melancholy task to hatch a set of little prisoners. How different was the case when we could anticipate the pleasure of rearing a family to all the joys of liberty! Men, it is true, now with officious care supply us with the necessary materials, and we make a very good nest; but I protest I had much rather be at the trouble of seeking them.

What pleasure have we experienced in plucking a bit of wool from a sheep's back, in searching for moss, in selecting the best feather where numbers were left to our choice, in stopping to rest on the top of a tree which commanded an extensive prospect, in joining a choir of songsters whom we accidentally met! But now our days pass with repeated sameness; variety, so necessary to give a relish to all enjoyment, is wanting. Instead of the songs of joy we formerly heard from every spray, our ears are constantly annoyed with the sound of mournful lamentations, transports of rage, or murmurs of discontent. Could we reconcile ourselves to the loss of liberty, it is impossible to be happy here unless we could harden our hearts to every sympathetic feeling."

"True," said his mate; "yet I am resolved to try what patience, resignation, and employment will effect, and hope, as our young ones will never know what liberty is, they will not pine as we do for it." Saying this, she picked up a straw, her mate followed the example, and they pursued their work.

At this instant a hen goldfinch brought forth her brood, who were full fledged. "Come, my nestlings," said she, " use your wings; I will teach you to fly in all directions." So saying, the little ones divided; one flew upwards, but emulous to outdo a little sparrow which was flying in the air above the aviary, he hit himself against the wires of the dome, and would have fallen to the bottom, but that he was stopped by one of the perches. As soon as he recovered,—

"Why cannot I soar as I see other birds do?" said he.

"Alas!" cried the mother, "we are in a place of confinement, we are shut up, and can never get out; but here is food in abundance, and every other necessary."

"Never get out!" exclaimed the whole brood; "then adieu to happiness!" She attempted to soothe them, but in vain.

The little redbreasts rejoiced in their liberty, and Dicky gave up the desire of living in the aviary, and wished to be gone.

"Stop," said the father, "let us first hear what those canary birds are saying."

The canary birds had almost completed their nest.

"How fortunate is our lot," said the hen bird, "in being placed in this aviary! How preferable is it to the small cage we built in last year!"

"Yes," replied her mate; "yet how comfortable was that in comparison with the still smaller ones in which we were once separately confined! For my part, I have no wish to fly abroad, for I should neither know what to do nor whither to go; and it shall be my endeavour to inspire my young ones with the same sentiments as I feel. Indeed, we owe the highest gratitude to those who make such kind provision for a set of foreigners who have no resources but their bounty, and my best lays shall be devoted to them. Nothing is wanting to complete the happiness of this place but to have other kinds of birds excluded. Poor creatures! it must be very mortifying to them to be shut up here, and see others of their kind enjoying full freedom. No wonder they are perpetually quarrelling; for my part, I sincerely pity them, and am ready to submit, from a principle of compassion, to the occasional insults and affronts I meet with."

"You now perceive, Dicky," said the cock redbreast, "that this place is not, as you supposed, the region of perfect happiness; you may also observe that it is not the abode of universal wretchedness. It is by no means desirable to be shut up for life, let the place of confinement be ever so splendid; but should it be our lot to be caught and imprisoned, which may possibly be the case, adopt the sentiments of the linnet and canary birds; employment will pass away many an hour that would be a heavy load if spent in grief and anxiety, and reflections on the blessings and comforts that are still in your power will lessen your regret for those which are lost. But come, pick up some of the seeds which are scattered on the outside of the aviary, for that is no robbery, and then I will show you another scene."

CHAPTER XVI.

THE OLD ROBINS TAKE LEAVE OF THEIR YOUNG ONES.

As soon as the redbreasts had regaled themselves with the superfluities of the feathered captives, they took their flight to a different part of the garden, in which was a collection of fowls and foreign birds. It consisted of a number of pens built round a grass-plot; in each was a pan of water, a sort of box containing a bed or nest, a trough for food, and a perch. In every pen was confined a pair of birds, and every pair was either of a different species, or distinguished for some beautiful variety either of form or plumage. The wooden bars which were put in the front were painted partly green and partly white, which dazzled the sight at the first glance, and so attracted the eyes that there was no seeing what was behind without going close up to the pens.

The little redbreasts knew not what sight to expect, and begged their parent to gratify their curiosity. "Well, follow me," said the father; "but I believe you must alight on the cross-bars, or you will not be able to examine the beauties of these fowls." They did so, and in the first pen was a pair of partridges. The size of these birds, so greatly exceeding their own, astonished them all; but notwithstanding this, the amiable Pecksy was quite interested by their modest, gentle appearance, and said she thought no one could ever wish to injure them.

"True, Pecksy," replied the father, "they have, from the harmlessness of their disposition, a natural claim to tenderness and compassion, and yet I believe there are few birds who meet with less; for I have observed that numbers share the same fate as the redstart which you saw die in the grove. I have myself seen many put to death in that manner. For a long time I was excessively puzzled to account for this fatality, and resolved, if possible, to gratify my curiosity. At length I saw a man kill two partridges and take them away. This very man had shown me great kindness, in feeding me when I first left my father's nest; so I had no apprehension of his doing me an injury, and resolved to follow him. When he arrived at his own house, I saw him deliver the victims of his cruelty to another person, who hung them up together by the legs, in a place which had a variety of other dead things in it, the sight of which shocked me exceedingly, and I could stay no longer. I therefore flew back to the field in which I had seen the murder committed, and in searching about found the nest belonging to the poor creatures, in which were several young ones just hatched, who in a short time would be starved to death! How dreadful is the fate of young animals that lose their parents before they are able to shift for themselves! and how grateful ought those to be to whom the blessings of parental instruction and assistance are continued!

"When the next morning arrived, I went again to see after the dead partridges, and found them hanging as before, and this was the case the day after; but the following morning I saw a boy stripping all their feathers off. As soon as he had completed this horrid operation, a woman took them, whom I ventured to follow, as the window of the place she entered stood open; where, to my astonishment, I beheld her twist their wings about and fasten them to their sides, then cross their legs upon their breasts, and run something quite through their bodies.

After this she put them before a place which glowed with a brightness something resembling the setting sun, which, on the woman's retiring, I approached, and found intolerably hot, I therefore made a hasty retreat; but resolving to know the end of the partridges, I kept hovering about the house; and at last, looking in at a window, I saw them, smoking hot, set before the man who murdered them, who was accompanied by several others, all of whom eyed them with as much delight as I have seen any of you display at the sight of the finest worm or insect that could be procured. In an instant after this the poor partridges were divided limb from limb, and each one of the party present had his share, till every bone was picked. There were other things devoured in the same manner; from which I learnt that men feed on birds and other animals, as we do on those little creatures which are destined for our sustenance, only they do not eat them alive."

"Pray, father," said Dicky, "do they eat redbreasts?" "I believe not," said he, "but I have reason to suppose they make many a meal of sparrows, for I have seen vast numbers of them killed."

At this instant their attention was attracted by one of the partridges in the pen, which thus addressed his mate:—"Well, my love, as there is no chance of our being set at liberty, I think we may as well prepare our nest, that you may deposit your eggs in it. The employment of hatching and raising your little ones will at least mitigate the wearisomeness of confinement, and I promise myself many happy days yet; for as we are so well fed and attended, I think we may form hopes that our offspring will also be provided for; and though they will not be at liberty to range about as we formerly did, they will avoid many of those terrors and anxieties to which our race are frequently exposed, at one season of the year in particular." "I am very ready to follow your advice," said the hen partridge, "and the business will soon be completed, for the nest is in a manner made for us, it only wants a little adjusting; I will therefore set about it immediately, and will no longer waste my hours in fruitless lamentations, since I am convinced that content will render every situation easy in which we can enjoy the company of our dearest friends, and obtain the necessaries of life." So saying, she retired into the place provided for the purpose on which she was now intent, and her mate followed, in order to lend her all the assistance in his power.

"I am very glad that my young ones have had

the opportunity of seeing such an example as this," said the hen redbreast. "You now understand what benefit it is to have a temper of resignation. More than half the evils of life, I am well convinced, arise from fretfulness and discontent; and would every one, like these partridges, try to make the best of their condition, we should seldom hear complaints, for there are much fewer real than imaginary misfortunes. But come, let us take a peep into the next pen."

They accordingly hopped to the next partition, in which were confined a pair of pencilled pheasants. Flapsy was quite delighted with the elegance of their form and the beauty of their plumage, and could have stayed the whole day looking at them; but as these birds were also tame and contented, nothing more could be learnt here than a confirmation of what the partridges had taught. Our travellers therefore proceeded still further, and found a pair of gold pheasants. Their splendid appearance struck the young redbreasts with astonishment, and raised such sentiments of respect, that they were even fearful of approaching birds which they esteemed as so much superior to themselves; but their father desiring they would never form a judgment of birds from a glittering outside, placed his family where they had an opportunity of observing that this splendid pair had but little intrinsic merit. They were proud of their fine plumage, and their chief employment was walking backwards and forwards to display it; and sometimes they endeavoured to push through the bars of their prison, that they might get abroad to show their rich plumage to the world, and exult over those who were in this respect inferior to them. "How hard," said one of them, "it is to be shut up here, where there are no other birds to admire us, and where we have no little ugly creatures to ridicule!"

"If such are your desires," said the hen redbreast "I am sure you are happier here than at liberty; for you would by your proud, affected airs excite the contempt of every bird which has right sentiments, and consequently meet with continual mortification, to which even the ugliest might contribute."

Pecksy desired to know if all fine birds were proud and affected. "By no means," replied her mother; "you observed the other pair of pheasants, who were, in my opinion, nearly equal to these for beauty and elegance. How easy and unassuming were they! and how much were their charms improved by the graces of humility! I often wonder that any bird should indulge itself in pride: what have such little creatures as we to boast of? The largest species amongst us is very inferior to many animals we see in the world, and man is lord over the greatest and strongest even of these. Nay, man himself has no cause to be proud, for he is subject to death as well as the meanest of creatures, as I have had opportunities of observing. But come, the day wears away; let us view the other parts of this enclosure."

On this the father conducted his family to a variety of pens, in which were different sorts of foreign birds, of whom he could give but little account, therefore would not suffer his young ones to stand gazing at them long, lest they should imbibe injurious notions of them; especially when he heard Dicky cry, as he left the pen, "I dare say that bird is a very cruel, voracious creature; I make no doubt but that he would eat us all, one after the other, if he could get at us."

"Take care, Dicky," said the father, "how you form an ill opinion of any one on slight grounds You cannot possibly tell what the character of this stork is merely from his appearance; you are a stranger to his language, and cannot see the disposition of his heart. If you give way to a suspicious temper, your own little breast will be in a state of constant perturbation; you will absolutely exclude yourself from the blessings of society, and will be shunned and despised by birds of every kind. The stork which you thus censure is far from deserving your ill opinion. He would do you no harm, and is remarkable for his filial affection. I saw him taken prisoner. He was carrying his aged father on his back, whom he had for a long time fed and comforted; the weight of this precious burden impeded his flight, and being at length weary with it, he descended to the ground to rest himself, when a cruel man, who was out on the business of bird-catching, threw a net over them, and then seized him by the neck. The poor old stork, who was before worn out with age and infirmities, unable to bear this calamity, fell from his back and instantly expired. The stork, after casting a look of anguish on his dear parent which I shall never forget, turned with fury on his persecutor, whom he beat with his wings with all the strength he had; but it was in vain to contend with a being so much more powerful than himself, and in spite of all his exertions he was conveyed to this place. But come, let us pick up a little refreshment, and then return to the orchard."

Saying this, he alighted on the ground, as did his mate and her family, where they met with a plentiful repast, in the provisions which had been accidentally scattered by the person whose employment it was to bring food for the inhabitants of the fowl-houses. When they had sufficiently regaled themselves, all parties gladly returned to the nest, and every heart rejoiced in the possession of liberty and peace.

For three successive days nothing remarkable happened either at Mr. Benson's or at the redbreasts' nest. The little family came to the breakfast-table, and Robin recovered from his accident, though not sufficiently to fly well; but Dicky, Flapsy, and Pecksy continued so healthy, and improved so fast, that they required no further care, and the third morning after their tour to the grove they did not commit the least error. When they retired from the parlour into the courtyard, to which Robin accompanied them, the father expressed great delight that they were at length able to shift for themselves.

And now a wonderful change took place in his own heart. That ardent affection for his young which had hitherto made him, for their sakes, patient of toil and fearless of danger, was on a sudden quenched; but, from the goodness of his disposition, he still felt a kind solicitude for their future welfare; and calling them around him, he thus addressed them:—

"You must be sensible, my dear young ones, that from the time you left the egg-shell to the present instant, both your mother and I have nourished you with the tenderest love. We have taught you all the arts of life which are necessary to procure you subsistence and preserve you from danger. We have shown you a variety of characters in the different classes of birds, and pointed out those which are to be shunned. You must now shift for yourselves; but, before we part, let me repeat my admonition, to use industry, avoid contention, cultivate peace, and be contented with your condition. Let none of your own species excel you in any amiable quality, for want of your endeavours to equal the best; and do your duty in every relation of life, as we have done ours by you. To the gay scenes of levity and dissipation prefer a calm retirement, for there is the greatest degree of happiness to be found. You, Robin, I would advise, on account of your infirmity, to attach yourself to the family where you have been so kindly cherished."

While he thus spake his mate stood by; who, finding the same change beginning to take place in her own breast, viewed her young ones with tender regret, and when he ceased cried out, "Adieu, ye dear objects of my late cares and solicitude! may ye never more stand in need of a mother's assistance! Though nature now dismisses me from the arduous task which I have long daily performed, I rejoice not, but would gladly continue my toil for the sake of its attendant pleasures. Oh, delightful sentiments of maternal love, how can I part with you? Let me, my nestlings, give you a last embrace." Then spreading her wings, she folded them successively to her bosom, and instantly recovered her tranquillity.

Each young one expressed his grateful thanks to both father and mother, and with these acknowledgments filial affection expired in their breasts, instead of which a respectful friendship succeeded. Thus was that tender tie dissolved which had hitherto bound this little family together; for the parents had performed their duty, and the young ones had no further need of their parental care.

The old redbreasts, having now only themselves to provide for, resolved to be no longer burthensome to their benefactors; and after pouring forth their gratitude in the most lively strains, they took their flight together, resolving never to separate. Every care now vanished, and their little hearts felt no sentiments but cheerfulness and joy. They ranged the fields and gardens, sipped at the coolest springs, and indulged themselves in the pleasures of society, joining their cheerful notes with those of other gay choristers which animate and heighten the delightful scenes of rural life.

The first morning that the old redbreasts were missing from Mrs. Benson's breakfast-table, Frederick and his sister were greatly alarmed for their safety; but their mamma said she was of opinion that they had left their nestlings, as it was the nature of animals in general to dismiss their young as soon as they were able to provide for themselves. "That is very strange," cried Harriet; "I wonder what would become of my brother and me, were you and papa to serve us so!"

"And is a boy of six, or a girl of eleven years old, capable of taking care of his or herself?" said her mamma. "No, my dear child, you have need of a much longer continuance of your parents' care than birds and other animals; and therefore God has ordained that parental affection, when once awakened, should always remain in the human breast, unless extinguished by the undutiful behaviour of a child."

"And shall we see the old redbreasts no more?" cried Frederick? "I do not know that you will," replied Mrs. Benson, "though it is not unlikely that they may visit us again in the winter; but let not their absence grieve you, my love, for I dare say they are safe and happy.'

At that instant the young ones arrived, and met with a very joyful reception. The amusement they afforded to Frederick soon reconciled him to the loss of their parents; but Harriet declared she could not help being sorry that they were gone. "I shall for the future, mamma," said she, "take notice of animals; for I have had much entertainment in observing the ways of these robins." "I highly approve of your resolution, my dear," said Mrs. Benson, "and hope the occasional instruction I have at different times given you has furnished you with general ideas respecting the proper treatment of animals. I will now inform you upon what principles the rules of conduct I prescribe to myself on this subject are founded.

"I consider that the same almighty and good Being who created mankind made all other living creatures likewise, and appointed them their different ranks in the creation, that they might form together a community, receiving and conferring reciprocal benefits. There is no doubt that the Almighty designed all beings for happiness, proportionable to the faculties He has endowed them with; and whoever wantonly destroys that happiness acts contrary to the will of his Maker.

"The world we live in seems to have been principally designed for the use and comfort of mankind, who, by the divine appointment, have dominion over the inferior creatures; in the exercise of which it is certainly their duty to imitate the Supreme Lord of the universe, by being merciful to the utmost of their power. They are endowed with reason, which enables them to discover the different natures of brutes, the faculties they possess, and how they may be made serviceable in the world; and as beasts cannot apply these faculties to their own use in so extensive a way, and numbers of them, being unable to provide for their own sustenance, are indebted to men for many of the necessaries of life, men have an undoubted right to their labour in return.

"Several other kinds of animals, which are sustained at the expense of mankind, cannot labour for them; from such they have a natural claim to whatever they can supply towards the food and raiment of their benefactors; and therefore, when we take the wool and milk of the flocks and herds, we take no more than our due, and what they can very well spare, as they seem to have an over-abundance given them, that they may be able to return their obligations to us.

"Some creatures have nothing to give us but their own bodies: these have been expressly destined by the Supreme Governor as food for mankind, and he has appointed an extraordinary increase of them for that very purpose, such an increase as would be very injurious to us if all were suffered to live. These we have an undoubted right to kill, but should make their short lives as comfortable as possible.

"Other creatures seem to be of no particular use to mankind, but as they serve to furnish our minds with contemplations on the wisdom, power, and goodness of God, and to exhilarate our spirits by their cheerfulness, they should not be wantonly killed, nor treated with the least degree of cruelty, but should be at full liberty to enjoy the blessings assigned them; unless they abound to such a degree as to become injurious, by devouring the food which is designed for man, or for animals more beneficial to him, whom it is his duty to protect.

"Some animals, such as wild beasts, serpents, &c., are in their nature ferocious, noxious, or venomous, and capable of injuring the health, or even of destroying the lives of men and other creatures of a higher rank than themselves; these, if they leave the secret abodes which are allotted them, and become offensive, certainly may with justice be killed.

"In a word, my dear, we should endeavour to regulate our regards according to the utility and necessities of every living creature with which we are anyways connected, and consequently should prefer the happiness of mankind to that of any animal whatever. Next to these (who, being partakers of the same nature with ourselves, are more properly our fellow-creatures) we should consider our cattle and domestic animals, and take care to supply every creature that is dependent on us with proper food, and keep it in its proper place; after their wants are supplied, we should extend our benevolence and compassion as far as possible to the inferior ranks of beings, and if nothing further is in our power, we should at least refrain from exercising cruelties on them. For my own part, I never willingly put to death, or cause to be put to death, any creature, but when there is a real necessity for it, and have my food dressed in a plain manner, that no more lives may be sacrificed for me than nature requires for my subsistence in that way which God has allotted me. But I fear I have tired you with my long lecture, so will now dismiss you."

While Mrs. Benson was giving these instructions to her daughter, Frederick diverted himself with the young redbreasts, who, having no kind parents now to admonish them, made a longer visit than usual; so that Mrs. Benson would have been obliged to drive them away, had not Pecksy, on seeing her move from her seat, recollected that she and her brother and sister had been guilty of an impropriety; she therefore reminded them that they should no longer intrude, and led the way out at the window; the others followed her, and Mrs. Benson gave permission to her children to take their morning's walk before they began their lessons.

CONCLUSION.

As the old robins, who were the hero and heroine of my tale, are made happy, it is time for me to put an end to it: but my young readers will doubtless wish to know the sequel of the history; I shall therefore inform them of it in as few words as possible.

Harriet followed her mamma's precepts and examples, and grew up a general benefactress to all people and all creatures with whom she was anyways connected.

Frederick was educated upon the same plan, and was never known to be cruel to animals, or to treat them with an improper degree of fondness; he was also remarkable for his benevolence, so as to deserve and obtain the character of a good man.

Lucy Jenkins was quite reformed by Mrs. Benson's lecture and her friend's example; but her brother continued his practice of exercising barbarities on a variety of unfortunate animals, till he went to school; where, having no opportunity of doing so, he gratified his malignant disposition on his schoolfellows, and made it his diversion to pull their hair, and pinch and tease the younger boys; and, by the time he became a man, had so hardened his heart that no kind of distress affected him, nor did he care for any person but himself; consequently he was despised by all with whom he had any intercourse. In this manner he lived for some years; at length, as he was inhumanly beating and spurring a fine horse merely because it did not go a faster pace than it was able to do, the poor creature, in its efforts to evade his blows, threw his barbarous rider, who was killed on the spot.

Farmer Wilson's prosperity increased with every succeeding year; and he acquired a plentiful fortune, with which he gave portions to each of his children as opportunities offered for settling them in the world; and he and his wife lived to a good old age, beloved and respected by all who knew them.

Mrs. Addis lost her parrot by the disorder with which it was attacked while Mrs. Benson was visiting at the house; and before she had recovered the shock of this misfortune, as she called it, her grief was renewed by the death of the old lapdog. Not long afterwards her monkey escaped to the top of a house, from whence he fell and broke his neck. The favourite cat went mad, and was obliged to be killed. In short, by a series of calamities all her dear darlings were successively destroyed. She supplied their places with new favourites, which gave her a great deal of fatigue and trouble.

In the meanwhile her children grew up, and having experienced no tenderness from her, they scarcely knew they had a mamma, nor did those who had the care of their education inculcate that her want of affection did not cancel their duty; they therefore treated her with the utmost neglect, and she had no friend left. In her old age, when she was no longer capable of amusing herself with cats, dogs, parrots, and monkeys, she became sensible of her errors, and wished for the comforts which other parents enjoyed: but it was now too late, and she ended her days in sorrow and regret.

This unfortunate lady had tenderness enough in her disposition for all the purposes of humanity, and had she placed it on proper objects, agreeably to Mrs. Benson's rule, she might have been, like her, a good wife, mother, friend, and mistress, consequently respectable and happy. But when a child Mrs. Addis was, under an idea of making her tender-hearted, permitted to lavish immoderate fondness on animals, the care of which engrossed her whole attention, and greatly interrupted her education; so that, instead of studying natural history and other useful things, her time was taken up with pampering and attending upon animals which she considered as the most important business in life.

Her children fell into faults of a different nature. Miss Addis being, as I observed in the former part of this history, left to the care of servants, grew up with very contracted notions. Amongst other prejudices, she imbibed that of being afraid of spiders, frogs, and other harmless things; and having been bitten by the monkey, and terrified by the cat when it went mad, she extended her fears to every kind of creature, and could not take a walk in the fields, or even in the street, without a thousand apprehensions. And at last her constitution, which from bad nursing had become very delicate, was still more weakened by her continual apprehensions; and a rat happening to run across the path as she was walking, she fell into fits, which afflicted her at intervals during the remainder of her life.

Edward Addis, as soon as he became sensible of his mother's foible, conceived an inveterate hatred to animals in general, which he regarded as his enemies, and thought he was avenging his own cause when he treated any with barbarity. Cats and dogs, in particular, he singled out as the objects of his revenge, because he considered them as his mother's greatest favourites; and many a one fell an innocent victim to his mistaken ideas.

The parent redbreasts visited their kind benefactors the next winter; but as they were flying along one day, they saw some crumbs of bread which had been scattered by Lucy Jenkins, who, as I observed before, had adopted the sentiments of her friend in respect to compassion to animals, and resolved to imitate her in every excellence. The redbreasts gratefully picked up the crumbs, and encouraged by the gentle invitation of her looks, determined to repeat their visits; which they accordingly did, and found such an ample supply that they thought it more advisable to go to her with their next brood than to be burthensome to their old benefactors, who had a great number of pensioners to support: but Frederick and Harriet Benson had frequently the pleasure of seeing them, and knew them from all their species by several peculiarities which so long an acquaintance had given them the opportunity of observing.

Robin, in pursuance of his father's advice, and agreeably to his own inclinations, attached himself to Mr. Benson's family, where he soon became a great favourite. He had before, under the conduct of his parents, made frequent excursions into the garden, and was, by their direction, enabled to get up into trees, but his wing never recovered sufficiently to enable him to take long flights; however, he found himself at liberty to do as he pleased, and during the summer months he commonly passed most of his time abroad, and roosted in trees, but visited the tea-table every morning; and there he usually met his sister Pecksy, who took up her abode in the orchard, where she enjoyed the friendship of her father and mother. Dicky and Flapsy, who thought their company too grave, flew giddily about together. In a short time they were both caught in a trap-cage, and put into the aviary which Dicky once longed to inhabit. Here they were at first very miserable; but after a while, recollecting their good parents' advice, and the example of the linnets and pheasants, they at length reconciled themselves to their lot, and each met with a mate, with whom they lived tolerably happy.

Happy would it be for the animal creation if every human being, like good Mrs. Benson, consulted the welfare of inferior creatures, and neither spoiled them by indulgence nor injured them by tyranny. Happy would mankind be if every one, like her, acted in conformity to the will of their Maker, by cultivating in their own minds, and those of their children, the divine principle of general benevolence.

From the foregoing examples I hope my young readers will select the best for their own imitation, and take warning by the rest; otherwise my STORY OF THE ROBINS will have been written in vain.

Frederick Warne & Co., Publishers.

WARNE'S
1s. 6d. "Birthday Series" of Gift-Books.

Pott 8vo, with Original Illustrations, cloth, gilt edges.

FRITZ; or, Experience Teacheth Wisdom.
THE LEONARDS; or, The Cobbler, The Clerk, &c.
CHILD'S FINGER-POST; or, Help for the Heedless.
BIRTHDAY STORIES FOR YOUNG PEOPLE.
BIRTHDAY TALES FOR YOUNG PEOPLE.
THE LOST HEIR; or, Truth and Falsehood.
HENRY BURTON; or, The Reward of Patience.
THE TORN BIBLE; or, Hubert's Best Friend.
COUSIN ANNIE; or, Heart and Hand.
MR. RUTHERFORD'S CHILDREN.
MAUD LATIMER; or, Patience and Impatience.
STORIES OF OLD. Bible Narratives for Young Children—Old Testament. By CAROLINE HADLEY.
Ditto   ditto   New Testament.
STORIES OF THE APOSTLES: Their Lives and Writings. By CAROLINE HADLEY.
SHORT TALES FOR SUNDAY READING.
PHILIP AND HIS GARDEN. By CHARLOTTE ELIZABETH.
STORIES OF THE BIBLE.   Ditto.
WILLIAM HENRY'S SCHOOLDAYS. By A. M. DIAZ.
WILLIAM HENRY AND HIS FRIENDS.   Ditto.
CHILDREN'S SAYINGS; or Early Life at Home. By CAROLINE HADLEY.
UNCLE JACK THE FAULT KILLER.
CHILDREN OF THE SUN.—Poems for the Young.

By CATHERINE D. BELL.

AN AUTUMN AT KARNFORD.
ARNOLD LEE; or, Rich and Poor Boys.
ALLEN AND HARRY; or, Set About it at Once.
GEORGIE AND LIZZIE; or, Self-Denial.
THE DOUGLAS FAMILY; or, Friendship.

WARNE'S
1s. 6d. "Fairy Series" of Gift-Books.

Pott 8vo, with Original Illustrations, cloth, gilt edges.

THE CHILDREN OF ELF LAND. New Fairy Tales. By F. J. PAULL.
STORIES FROM FAMOUS BALLADS. By GRACE GREENWOOD.
THE NEW YEAR'S BARGAIN. By SUSAN COOLIDGE.
THE DAISY AND HER FRIENDS. By Mrs. BRODERIP.
DREAMLAND; or, Children's Fairy Tales.
TALES FOR THE YOUNG. By HANS ANDERSEN.
NURSERY TALES. A New Version. By Mrs. VALENTINE.

Bedford Street, Strand.

Frederick Warne & Co., Publishers.

Warne's One Shilling "Round the Globe" Library.

Large fcap. 8vo, cloth gilt, with Coloured Frontispiece & Woodcut Illustrations.

TILLY TRICKETT; or, Try. By M. KEARY.
ALEC DEVLIN; or, Choose Wisely. By Mrs. F. AYLMER.
MANOR HOUSE EXHIBITION AND THE DARRELL MUSEUM.
THE CHILDREN'S GARDEN, and What they Made of it.
WILLIE HERBERT AND HIS SIX LITTLE FRIENDS.
OLD GINGERBREAD AND THE SCHOOLBOYS.
THE SEVEN KINGS OF ROME, AND THE STORY OF POMPEII.
THE EARTH WE LIVE ON.
THE ITALIAN BOY, AND INDUSTRIAL MEN OF NOTE.
HOME TEACHINGS IN SCIENCE.
CHAT IN THE PLAYROOM, AND LIFE AT A FARMHOUSE.
OUR PONDS AND OUR FIELDS, &c.
BRAVE BOBBY, PETER AND HIS PONY, &c.
THE PEASANTS OF THE ALPS, &c.
FRANCES MEADOWS, TRAITS OF CHARACTER, &c.
UNCLE JOHN'S ADVENTURES AND TRAVELS.
CASPAR.
CARL KRINKEN. } By the Author of "The Wide, Wide World."
FRANK RUSSELL; or, Living for an Object.
TOM BUTLER'S TROUBLE. A Cottage Story.
LIZZIE JOHNSON; or, Mutual Help.
MR. RUTHERFORD'S CHILDREN. 1st series. Ditto, 2nd series.
THE CHILDREN'S HARP; or, Select Poetry.
CHARLIE CLEMENT; or, The Boy Friend.
A QUEEN. A Story for Girls.
RUTH CLAYTON. A Book for Girls.
NELLIE GRAY; or, Ups and Downs of Life.
CLARA WOODWARD; and her Day Dreams.
SUSAN GRAY. By Mrs. SHERWOOD.
THE LITTLE MINER; or, Truth and Honesty.
EASY RHYMES AND SIMPLE POEMS.
MY EARNINGS; or, Ann Ellison's Life.
BABES IN THE BASKET. By AUNT FRIENDLY.
BASKET OF FLOWERS. Revised Edition.
SAM; or, A Good Name. By M. KEARY.
EDITH AND MARY; or, Holly Farm.
WILLIE'S BIRTHDAY.
THE SILVER TRUMPET.
WILLIE'S REST. A Sunday Story.
UNICA. A Story for Sunday.
STORY BOOK OF COUNTRY SCENES.
TWILIGHT STORIES AT OVERBURY FARM.
GOLD SEEKERS AND BREAD WINNERS.
STUYVESANT; or, Home Adventures.
CAROLINE; or, The Henrys. } By JACOB ABBOTT.
AGNES; or, Summer on the Hills.
MARY ELTON.
PRIDE AND PRINCIPLE.
THEODORA'S CHILDHOOD.
MRS. GORDON'S HOUSEHOLD.
LITTLE NETTIE; or, Home Sunshine.
ROBERT DAWSON; or, The Brave Spirit.
THE DAIRYMAN'S DAUGHTER.
JANE HUDSON; or, The Secret of Getting on.
LITTLE JOSEY; or, Try and Succeed.
THE YOUNG COTTAGER.
MASTER GREGORY'S CUNNING.

Bedford Street, Strand.

Frederick Warne & Co., Publishers.

The Star Library.

In this Series, from time to time will be issued a very popular edition of well-known Books, many of them copyright, produced with the sanction of the proprietors of the "Golden Ladder" Series, and published at prices, united with style and completeness, hitherto unequalled.

One Shilling Volumes, stiff Picture Wrapper.

DAISY. By ELIZABETH WETHERELL.
DAISY IN THE FIELD. By ELIZABETH WETHERELL.
NETTIE'S MISSION. By ALICE GRAY.
STEPPING HEAVENWARD. By E. PRENTISS.
WILLOW BROOK. By ELIZABETH WETHERELL.
SCEPTRES AND CROWNS, AND THE FLAG OF TRUCE. By ELIZABETH WETHERELL.
DUNALLAN. By GRACE KENNEDY.
FATHER CLEMENT. By GRACE KENNEDY.

Eighteenpenny Volumes.

WIDE, WIDE WORLD. By ELIZABETH WETHERELL.
QUEECHY. By ELIZABETH WETHERELL.
MELBOURNE HOUSE. By ELIZABETH WETHERELL.
DRAYTON HALL. By ALICE GRAY.
SAY AND SEAL. By ELIZABETH WETHERELL.

Bedford Street, Strand.

www.ingramcontent.com/pod-product-compliance
Lightning Source LLC
Chambersburg PA
CBHW021839230426
43669CB00008B/1013